K. Blaum
F. Herfurth

Trapped Charged Particles and Fundamental Interactions

T0224105

 Springer

Klaus Blaum
Universität Mainz
Inst. Physik
Staudinger Weg 7
55099 Mainz
Germany
blaumk@uni-mainz.de

Frank Herfurth
GSI Darmstadt
Planckstr. 11
64291 Darmstadt
Germany
F.Herfurth@gsi.de

Blaum, K. and Herfurth, F. *Trapped Charged Particles and Fundamental Interactions*, Lect. Notes Phys. 749 (Springer, Berlin Heidelberg 2008), DOI 10.1007/ 978-3-540-77817-2

ISBN 978-3-642-09660-0   e-ISBN 978-3-540-77817-2

DOI 10.1007/978-3-540-77817-2

Lecture Notes in Physics ISSN: 0075-8450

Cover design: eStudio Calamar S.L., F. Steinen-Broo, Pau/Girona, Spain

Printed on acid-free paper

9 8 7 6 5 4 3 2 1

springer.com

Lecture Notes in Physics

The Lecture Notes in Physics

The series Lecture Notes in Physics (LNP), founded in 1969, reports new developments in physics research and teaching – quickly and informally, but with a high quality and the explicit aim to summarize and communicate current knowledge in an accessible way. Books published in this series are conceived as bridging material between advanced graduate textbooks and the forefront of research and to serve three purposes:

- to be a compact and modern up-to-date source of reference on a well-defined topic

- to serve as an accessible introduction to the field to postgraduate students and nonspecialist researchers from related areas

- to be a source of advanced teaching material for specialized seminars, courses and schools

Both monographs and multi-author volumes will be considered for publication. Edited volumes should, however, consist of a very limited number of contributions only. Proceedings will not be considered for LNP.

Volumes published in LNP are disseminated both in print and in electronic formats, the electronic archive being available at springerlink.com. The series content is indexed, abstracted and referenced by many abstracting and information services, bibliographic networks, subscription agencies, library networks, and consortia.

Proposals should be sent to a member of the Editorial Board, or directly to the managing editor at Springer:

Christian Caron
Springer Heidelberg
Physics Editorial Department I
Tiergartenstrasse 17
69121 Heidelberg / Germany
christian.caron@springer.com

Preface

Storage and cooling techniques for charged particles gain more and more importance in various areas of modern science. They developed into a universal tool especially when used for precision measurements. For this purpose, there are mainly two types of ion traps in use: radio frequency quadrupole (Paul) traps which use a time-varying quadrupolar electric field applied to the electrodes for confinement and Penning traps where a superposition of a homogeneous magnetic field with a weak electrostatic quadrupolar field is used. Already the very first experiments in ion traps, performed by their inventors Wolfgang Paul and Hans Dehmelt, paved the way for astonishingly precise measurements of fundamental quantities like the electron and positron g-factors and the fine-structure constant α. Their work was honored with the Nobel Prize in physics for "the development of the ion trap technique" in 1989. Since then many experimental physicists worldwide have been using and developing different kinds of ion traps. Today, ion traps are applied widely for instance in mass spectrometry, metrology, plasma physics, molecular and cluster physics, quantum computing, atomic and nuclear physics as well as in chemistry.

Precise investigations are able to link measurable quantities to fundamental aspects of physics. Due to the achievable precision, ion traps have been used for this subject and attracted a conference series "Trapped Charged Particles and Fundamental Interactions." Along the main idea of that conference we organized a Heraeus Winter School that took place in Hirschegg, Austria, in spring 2006. Inspired by the success and the interest from the students we planned a book that should contain the key components of the school: interesting, introductory and up-to-date lectures connected with ion traps.

The volume starts with a theoretical introduction into precision tests of electroweak theory with experiments performed at low energy. The present searches for physics beyond the Standard Model, that are at the same time precision test of the Standard Model itself, are presented and put into their theoretical context. This includes the search for time reversal violation in nuclear beta decay or via permanent electric dipole moments and the experiments about nature and mass of the neutrinos. The second introductory part of the volume deals with the basics of ion trapping and cooling. The principles of operation of the two trap types, Penning traps and Paul traps, are explained and technical details that are important for using them are given. Since manipulation and cooling are essential for many applications of these devices the most important detection and cooling techniques are presented.

In the second part a practical question that arises sooner or later to all experimentalists dealing with ion traps is discussed: What does the stored particle do in the trap and how can one simulate its motion during storage? The first chapter of this part introduces different methods to model ion dynamics and describes their implementation. In the second chapter the cooling of ions by collisions with buffer gas, one of the most important and most universal tools in ion manipulation in traps is dealt with in detail, the cooling of ions by collisions with buffer gas atoms is discussed. Different numerical approaches and their results for a few existing devices are introduced to the reader.

The third part of the volume is devoted to two applications discussed in depth. The first example is the use of highly charged ions to obtain high-precision mass values. The method is reviewed and examples related to fundamental questions in modern physics are presented. The second application deals with the storage and investigation of charged antiprotons, which not only is one of the prerequisites for the production of antihydrogen, but also allows in several ways to test one of the most fundamental symmetries, the combined charge, parity and time reversal symmetry, CPT. Ongoing and planned experiments are described as well.

We hope that this volume gives the experienced reader an in-depth view of some aspects of trapped charged particles that he/she might only have touched by now. But most importantly, this book shall give a newcomer to the field some feeling, thoughts and maybe also an introduction about the most interesting aspects at the borderline between modern trapping techniques and fundamental physics.

We would like to thank all authors for their effort to create a concise and yet recent picture of what physics is driving us. Special thanks go to H.-J. Kluge for the idea of the school and his help to organize it; to Christian Caron, our contact at Springer for the easy and smooth connection; and to the Heraeus Foundation and GSI Darmstadt for the kind financial support of the winter school on trapped charged particles.

Darmstadt, 2007

Klaus Blaum
Frank Herfurth

Contents

Low-Energy Precision Tests of Electroweak Theory

R.G.E. Timmermans

1 Beyond the Standard Model of Particle Physics

1.1 Our Pride and Joy: The Standard Model

In 2008, the Large Hadron Collider (LHC) will start up at CERN and open a new energy frontier in particle physics. Its main mission will be to find the Higgs boson predicted by the Standard Model (SM) of particle physics and to determine its properties. The Higgs sector, responsible for electroweak symmetry breaking, is the last missing piece of the SM, which otherwise has been confirmed in spectacular detail. Despite its many successes, however, there are strong reasons to believe that there is a theory *beyond* the SM, implying the existence of new particles that may manifest themselves at the scales accessible by the LHC. Indications for such new physics come primarily from cosmology (dark matter, baryogenesis, possibly inflation), from nonaccelerator experiments (in particular, the recent ground-breaking discovery of neutrino oscillations), and also from theoretical considerations. Numerous extensions of the SM have been proposed based on concepts like gauge unification, new strong interactions, supersymmetry (SUSY), or extra dimensions. At present, the experimental facts do not favor any particular model, and surprises are far from excluded.

The SM of particle physics is the culmination of the enormous progress made in the twentieth century to uncover the laws of physics at subatomic scales. It has many good features that one should keep in mind when thinking about improvements of the SM. The SM provides a *unified* framework for understanding three of the four known forces in the present universe: the electromagnetic and the strong

R.G.E. Timmermans
Theory Group, KVI, University of Groningen, Zernikelaan 25, 9747 AA Groningen
The Netherlands

Lectures given at the WE-Heraeus-Winterschool on "Trapped Charged Particles for Fundamental Interactions," Hirschegg, Germany, March 30–April 8, 2006.

Timmermans, R.G.E.: *Low-Energy Precision Tests of Electroweak Theory.* Lect. Notes Phys. **749**, 1–30 (2008)
DOI 10.1007/978-3-540-77817-2_1

and weak nuclear forces. The unification of fundamental forces, which started with Ampère, Faraday, and especially Maxwell, is high on the agenda of physics. The SM is a quantum field theory, which is the only known way to combine the special theory of relativity with the principles of quantum mechanics. At the foundation of the SM, in a sense an elaborate copycat of our prototype quantum field theory, QED, is a simple and elegant gauge principle, based on the group $SU(3) \times SU(2) \times U(1)$, which implements the idea that local symmetries result in forces. Thus, the SM unifies electromagnetism and weak interaction into the electroweak theory, and since the strong interaction is also described by a gauge theory, the road appears open for a *true* unification, in one gauge group with one universal gauge coupling, of the three forces at a scale of the order 10^{16} GeV within the paradigm of grand unification theories (GUTs). GUTs elegantly put quarks and leptons into single group representations and make important predictions such as proton decay, the strength of gauge couplings, and detailed properties of the quark and lepton mass matrices.

Most important, of course, is that the SM is astonishingly successful in quantitatively describing Nature: Its predictions have been verified and all experiments (except neutrino oscillations) are consistent with it. The new particles that it predicted (except the Higgs boson) have all been found. The theory of strong interaction, QCD, explains quantitatively the physics of deep-inelastic scattering and scaling violations, heavy-quark systems, and chiral dynamics. Despite the progress in lattice QCD, it remains hard to deal with QCD in the nonperturbative low-energy regime; nevertheless, it is clear that QCD provides the microscopic basis for all of hadronic and nuclear physics. The electroweak sector of the SM explains such varied phenomena as parity violation, neutral currents, quark-flavor mixing, CP violation, lepton universality, and muon and nuclear β-decay. Agreement with experiment in many cases requires calculations that include quantum loop effects. Finally, the SM provides the foundation for explaining many astrophysical and, combined with the general theory of relativity, many cosmological observations, such as primordial nucleosynthesis, the cosmic microwave background, weak interactions in stars, solar fusion, supernovae and neutron stars, and cosmic rays.

1.2 The Standard Model and Its flaws

In the SM, the Higgs sector [1] is needed to provide electroweak symmetry breaking. Without the Higgs boson, which acts as an ultraviolet cutoff, the SM does not make sense as a renormalizable quantum field theory. However, an important theoretical reason for believing that the SM is incomplete is that the SM with one Higgs boson is "unnatural," in the technical sense that (within the framework of unified gauge theories) it requires an extreme fine-tuning at the very high energy where the different forces become unified. It is therefore believed that the mechanism for electroweak symmetry breaking could well be more complicated than in the SM. The LHC experiments are designed to reveal the existence and properties of the Higgs particle(s). To make progress, the answer from the LHC about electroweak symmetry breaking is eagerly awaited.

Moreover, the Higgs sector is expected to provide some clues to other mysterious features of the SM as well. In the SM all known particles (the vector bosons and the quarks and charged leptons) get their mass exclusively via the Higgs system. Therefore, the Higgs boson is the only part of the SM that "knows about" the difference between the fermion families. Also parity violation could be said to follow from this: The only alternative way to give mass to the fermions is "by hand," i.e., by simply adding a mass term to the Lagrangian. This mass term, however, must be gauge invariant by itself, and hence left- and right-handed fermions must couple in the same way to the vector bosons, implying that parity is conserved. Because the Higgs field gives mass to the quarks, there is also a connection to CP violation, which in the SM requires precisely (at least) three families of quarks with nonzero mass. Finally, the SM with one Higgs boson is at odds with the observed small value of the cosmological constant; in fact, it predicts a universe curled up to the size of a football... [1].

There are other, more indirect, reasons to suspect that the SM is not complete. The SM has a large number of parameters, the values of which are a priori arbitrary and have to be determined from experimental data. In total, the SM has 19 such parameters: There are six quark masses, three masses for the charged leptons, and three gauge coupling constants; quark-flavor mixing is parametrized by the Cabibbo–Kobayashi–Maskawa (CKM) matrix in terms of three mixing angles and one phase that results in CP violation; from the Higgs sector we get the mass of the Higgs boson and the scale of electroweak symmetry breaking; finally, we should also count the QCD vacuum angle θ. One might hope that in a more complete theory the values of all, or at least some, of the SM parameters could be understood [2].

In the SM, the neutrinos are assumed to be massless. Now that neutrino oscillations are established, the question arises how to extend the SM to include the neutrino mass. If neutrinos are Dirac fermions, this is possible in a straightforward way, at the cost, however, of seven more parameters: three masses and three mixing angles and one phase to parametrize the flavor-mixing matrix. If neutrinos are Majorana fermions, on the other hand, this matrix will contain two additional phases, for a grand total of 28 parameters for the SM with massive Majorana neutrinos. Other questions that are left open by the SM include the following: Why the specific gauge group $SU(3) \times SU(2) \times U(1)$? Why are there three families of fermions? What is the deeper reason for the fermion mass hierarchy? What is the origin of parity violation and CP violation in the weak interaction? While we believe that it will be possible to find out the mechanism of electroweak symmetry breaking, be it the SM Higgs boson or something else, it may be that we never get an answer to some of these puzzles.

1.3 The Low-Energy Frontier

In particle physics, in our quest to find new particles, forces, and symmetries in order to extend the SM to a more complete theory, two, largely complementary, experimental approaches can be distinguished:

1. The traditional approach is through collider experiments at high energy. Clearly, this is the indicated, historically very successful, route to make *direct* observations of new particles. The high-energy frontier has now moved to the LHC at CERN, which is direly needed to get an answer on the structure of the Higgs sector, on which so much else depends, to make progress in particle physics. A key task for theory in this high-energy arena will be to distinguish any novel signal from a large background of known SM origin. It will be crucial to perform detailed comparisons of precise theoretical predictions with accurate data. Such predictions require the inclusion of higher-order quantum corrections, which in some cases can be sensitive to the existence of not-yet-discovered heavy particles, even if these are not produced directly. Moreover, since most reactions of interest involve the production of unstable particles that decay subsequently, the theoretical predictions must account for processes with a large number of final-state particles. Theorists in this field have at their disposal powerful tools for the automatization of calculations, such as the FORM computer algebra program and accurate and flexible Monte Carlo tools.

2. Apart from the big collider experiments, many possibilities exist to test the SM in small(er)-scale precision experiments at low(er) energy. Also here one can search directly for new particles, for instance, axions or dark-matter particles in underground laboratories. But *indirect* searches for new physics are also possible by the *precision measurement* of observables that can be calculated to (sufficiently) high precision within the SM. A significant deviation between the measurement and the calculation in that case establishes beyond doubt new physics. A verification of the SM predictions can be used to set limits on parameters in speculative SM extensions. This approach can also be used to determine accurate values for fundamental constants and SM parameters. Experiments in this arena are usually manifold, relatively cheap, and have a relatively quick turnaround time. In some cases, when they are really almost "tabletop," they exploit the sharpest weapons of atomic physics, such as lasers and particle traps.

The landscape of such low-energy precision measurements is vast and ever increasing. It is characterized by well-motivated experiments at dedicated small-scale infrastructures. The discovery potential of such experiments is robust and they are (designed to be) complementary to high-energy direct searches: For the specific observables addressed in these experiments, the discovery potential exceeds that of direct searches. A state-of-the-art overview of the field of low-energy precision physics can be found in the recent road map for the European strategy for nuclear [3] and for particle physics [4].

In this chapter, I will discuss just a few examples of low-energy precision experiments. My goals are quite modest: First, I would like to show you some nice physics. To apply some focus, I have chosen topics that all involve the *violation of discrete symmetries*. Second, I would like to convince you that such experiments have the potential to make major contributions to particle physics, although this may be superfluous for the audience of trap experts. Third, I will even mention some experiments that involve traps, even ion traps! Let me first introduce the topic of discrete symmetries, before discussing in detail examples for each of the three:

(a) The search for time-reversal violation.
(b) Is the neutrino its own antiparticle?
(c) Atomic parity violation and the Weinberg angle.

The first topic, which includes nuclear β-decay and electric dipole moments, and the third topic have obvious relevance for atom and ion traps. For the second topic, neutrino physics, there is less excuse; however, it is an important field in particle physics in which there have been major developments in recent years. Moreover, the accurate determination of nuclear masses and Q-values with traps is of significance here as well. And who knows, one of you may come up with an idea to measure the neutrino mass in a tabletop experiment with a trap... I should mention that for this write-up I have not made a serious attempt to be complete in giving references, and limited myself to a number of historically important research papers and to important recent results. I did try to include useful and accessible review papers or books, where available, from which the literature can be traced.

1.4 C, P, T, and All That: The Violation of Discrete Symmetries

The violation of the discrete symmetries charge conjugation (C), parity (P), and time reversal (T) is a powerful probe of new physics beyond the electroweak scale. It is important to realize that C, P, and T were "theoretical discoveries." Parity was introduced by Wigner in 1927/1928, after the discovery of quantum mechanics. He explained some mysterious regularities in atomic spectra as the result of parity selection rules in electromagnetic transitions. He also pointed out that parity has no classical counterpart; the parity operator, $U_P\psi(\vec{r},t) = \eta_P\psi(-\vec{r},t)$, and its associated conserved quantum number are important in quantum mechanics because of its formulation in a vector space and the superposition principle that results therefrom. Time reversal was also introduced by Wigner in 1932. The time-reversal operation is peculiar because it relates initial and final states in quantum mechanics, and hence no quantum number gets associated with time reversal. Technically, time reversal is implemented as an anti-unitary operator that involves complex conjugation, viz., $A_T\psi(\vec{r},t) = \eta_T\psi^*(\vec{r},-t)$. Charge conjugation, of course, had to await the formulation of the Dirac equation and following that the development of quantum field theory with the concept of antimatter. It was introduced by the Dutch physicist Kramers in 1937. It is more properly called particle–antiparticle conjugation.

The discrete symmetries C, P, and T were formalized in the framework of quantum field theory (QED), a development which culminated around 1955 in the celebrated CPT theorem of Pauli and others: *The combined* CPT *operation is a symmetry of any local Lorentz-invariant quantum field theory.* This is a very strong statement: It is impossible to build a Lagrangian out of Lorentz covariant, local quantum fields which is not CPT invariant. This holds not only for the SM, but for all gauge theories, including for instance supersymmetric theories. In the context of theories of quantum gravity, such as string theory, which has nonlocal aspects, the validity of the CPT theorem becomes an issue. Lorentz invariance can then be spontaneously broken, leading to observable frame-dependent effects. In fact, numerous low-energy precision experiments with atomic physics techniques search for violation of Lorentz and CPT symmetry, for instance laser spectroscopy of antihydrogen at the CERN Antiproton Decelerator. For the purpose of this chapter, however, we will assume CPT invariance, thereby implying that CP violation is equivalent to T violation.

The rest, as they say, is history. Right after the formulation of the CPT theorem, it was discovered in 1957 that parity (and charge conjugation) was violated 100% in the weak interaction, following the seminal theoretical work of Lee and Yang. It became clear that assuming discrete symmetries to be obeyed had been a theoretical prejudice, and that the validity of these, or any other, symmetries rests on experiment. In 1964, CP was found to be broken (slightly) in the oscillations of the neutral-kaon system. In 1998, the CPLEAR experiment at CERN claimed to have found T violation in neutral-kaon oscillations. Direct CP violation in the decays of neutral kaons was observed in 1999, and in 2001 CP violated was measured in the decay of B^0-mesons. The present status is that all the discrete symmetries C, P, T, and all their combinations, except CPT, are broken in Nature. So far, all these experimental findings are accounted for in the SM, where the weak interaction is purely "left-handed," and the phase of the CKM matrix economically parametrizes CP violation. The question about the deeper origin of the violation of the discrete symmetries, however, is wide open.

2 The Search for Time-Reversal Violation

2.1 Introduction

The origin of the combined CP (or equivalently T) violation is an outstanding question in particle physics. It has important implications for the observed cosmological asymmetry between matter and antimatter, which, following a profound early idea due to Sakharov [5], can be due to CP-breaking forces in the early universe. To explain the magnitude and sign of the asymmetry, known accurately from cosmic microwave background measurements (WMAP), CP violation within the SM appears to be far insufficient. Therefore new sources of CP violation are anticipated, which may shed light on the quark and lepton family structure and its relation to the as-yet-undiscovered Higgs system. CP violation will be addressed by numerous experiments at the LHC, underground laboratories, and by high-precision smaller-scale experiments that search for T violation.

The SM has a very peculiar flavor structure: Because there are three families of quarks and because the mass eigenstates are different from the weak eigenstates, the resulting CKM quark-mixing matrix contains one phase, δ_{CKM}, that gives rise to CP violation. This implies that in the SM, all CP-odd effects due to δ_{CKM} must involve all three families of quarks. For example, the $\Delta S = 2$ strangeness oscillations from $d\bar{s}$ to $s\bar{d}$ in the neutral-kaon system are due to the box diagrams with double $W^{\pm}$ exchange and virtual u, c, and t quarks; direct CP violation in kaon decay results from the famous penguin diagrams with a loop containing u, c, and t quarks. In fact, Jarlskog [6] has shown that all CP-odd effects in the SM are proportional to

$$J \sim \sin^2 \theta_1 \sin \theta_2 \sin \theta_3 \sin \delta_{CKM} ,$$

where θ_i ($i = 1, 2, 3$) are the quark-mixing angles, so CP violation requires three families with nonzero mass and mixing and additionally a nonzero δ_{CKM}.

This has important consequences for low-energy searches of CP or T violation, which involve only the first generation of up- and down-quarks. In flavor-conserving nonleptonic interactions, there is *no* T violation due to δ_{CKM} in first-order weak interactions. We can therefore give a naive, back-of-the-envelope, estimate of d_n, the electric dipole moment of the neutron (due to δ_{CKM}), which will be discussed in detail below. It has to be second order in the weak interaction, proportional to J, and we give it natural units e/M, where M is the nucleon mass. Thus, we guesstimate

$$(d_n)_{SM} \simeq 10^{-6} 10^{-6} s_1^2 s_2 s_3 s_\delta \times e/M \sim 10^{-30} \; ecm ,$$

where we write $s_i = \sin \theta_i$ and $s_\delta = \sin \delta_{CKM}$. This is some 4–5 orders of magnitude below the present experiment limit, so any experimental signal in the near future implies new physics. Similarly, in semileptonic processes, there is *no* T violation due to δ_{CKM} in first-order weak interactions. For the T-odd triple correlation D in nuclear β-decay, also discussed below, relative to let us say the T-even correlation a, we expect therefore

$$(D/a)_{SM} \simeq 10^{-6} s_1^2 s_2 s_3 s_\delta \sim 10^{-10} ,$$

again orders of magnitude below the present experimental limit.

CP violation is also present in the strong-interaction part of the SM, QCD. The symmetries of QCD allow a P-odd and T-odd interaction, parametrized by the QCD vacuum angle θ. Such a term results in a nonzero value for the neutron EDM. The present experimental limit on d_n implies that, for mysterious reasons, the value of θ is apparently unnaturally small. An important question is whether, when nonzero hadronic EDMs are measured, it can be decided if they are due to a nonzero value of θ, or to new physics, e.g., SUSY.

2.2 T Violation in Nuclear β-Decay

We have seen that the low-energy search for T violation provides us with a large window on new physics. But how to find T violation? It turns out that high-precision experiments that test time-reversal invariance are quite rare. One could, for instance, test S-matrix reciprocity: $S_{f,i} = S_{-i,-f}$, that is, one compares a reaction to its inverse. A well-known example is the equality of polarization and analyzing power in proton–proton elastic scattering. But since *two* experiments are involved, it will be impossible to reduce the systematic errors to the required level for a high-precision test. This is in contrast to parity, where one can make a precision test with just *one* scattering experiment, e.g., check the absence of polarization in a certain direction.

A second option for a precision T-violation experiment is to search for a nonzero value of a T-odd operator in a nondegenerate state. The prime example here is the electric dipole moment, which is discussed in Sect. 2.3. Another possibility is to search for a nonzero value of a T-odd observable after a weak decay. One has to be aware, however, that this can also be due to final-state interactions (FSI) that are perfectly well allowed in the SM. Suppose H_w is the weak Hamiltonian and S_0 is the S-matrix due to the FSI. T invariance together with unitarity implies, schematically,

$$\langle -f|H_w|-i\rangle = \sum_{f'}\langle f|S_0|f'\rangle\langle f'|H_w|i\rangle = \langle f|H_w|i\rangle \, ,$$

where the last equality, however, only holds when $S_0 = 1$, i.e., when FSI can be neglected. The fact that (T-even) FSI lead to a nonzero value for a T-odd observable after a weak decay is sometimes called "T-violation mimicry." It implies that this kind of experiment can be used to search for T violation only up to the point that the FSI are reached. Theoretical calculations of the pertinent FSI are therefore mandatory before embarking on an experiment.

The example relevant for us is the *triple correlations* in nuclear β-decay. The differential decay distribution can be written schematically as follows:

$$\frac{d^2\Gamma}{d\Omega_e d\Omega_\nu} \sim 1 + a\hat{p}\cdot\hat{q} + b\frac{m_e}{E_e}$$
$$+\langle\vec{J}\rangle\cdot[A\hat{p}+B\hat{q}+D\hat{p}\times\hat{q}]$$
$$+\langle\vec{\sigma}\rangle\cdot\left[G\hat{p}+N\langle\vec{J}\rangle+R\langle\vec{J}\rangle\times\hat{p}\right] \, ,$$

where we left out some terms that do not concern us here, and we define $\hat{p} = \vec{p}/E_e$ and $\hat{q} = \vec{q}/E_\nu$, where $\vec{p}$ and $\vec{q}$ are the momenta of the β-particle and neutrino, respectively; $\vec{J}$ and $\vec{\sigma}$ are the spins of the parent nucleus and the β-particle. Since both spin and momentum vectors are odd under T, nonzero values for the triple correlations D and R are evidence for T violation, provided that the FSI can be ignored. The FSI in this case are of electromagnetic origin, mostly the Coulomb interaction between the outgoing β-particle and the daughter nucleus. They can be calculated accurately.

The discovery potential of the different β-decay correlations can be analyzed in a general, model-independent, way starting from a general charged-current interaction that includes the possible Lorentz covariants with vector (V), axial-vector (A), tensor (T), scalar (S), and pseudoscalar (P) structure, see, e.g., [7]. In the low-energy limit, when the momentum transfer to the $W^\pm$ boson can be neglected compared to its mass, the SM reduces to the left-handed $V - A$ current–current interaction, but new physics could manifest itself as additional $V + A$, S, or T contributions (at low energies possible pseudoscalar terms are negligible). Such contributions would be induced, e.g., by gauge bosons with coupling to right-handed fermions, charged scalar Higgs bosons, or leptoquarks. The triple correlation multiplied by the D-coefficient is sensitive to new P-odd, T-even interactions with vector and axial-vector currents, while the triple correlation R probes P-odd, T-odd S or T interactions. D is in particular sensitive to T-odd interactions involving leptoquarks; measurements of R receive tough competition from electric dipole experiments, which at present provide more stringent limits. Both for D and R, the values predicted by the SM are orders of magnitude below the present experimental limits and the size of FSI, for the reasons given in Sect. 2.1.

Nuclear β-decay experiments, of course, have a long history. They were instrumental in establishing the $V - A$ structure of the weak interaction. To measure the D correlation mixed Fermi-Gamow-Teller decays have to be considered. It has been measured by the Princeton group in ^{19}Ne decay and, more recently, in neutron decay at NIST and ILL. The R correlation has been measured in ^{19}Ne decay at Princeton and in ^{8}Li decay at PSI. In recent years, the β-decay field has gone through a renaissance, with the advent of atom and ion traps and the corresponding progress in laser technology to cool and trap specific radioactive isotopes in sufficient amounts [8]. In this way, new levels of precision can be reached, and moreover the recoiling daughter nucleus can be detected, so that new correlations can be addressed (the recoil energies are typically in the tens of eV range).

A number of high-precision results for the β-neutrino correlation a in atom traps have already been obtained, in particular for the mixed decay of ^{21}Na at Berkeley [9] and for the Fermi decay of ^{38m}K at TRIUMF [10]. The Berkeley measurement disagrees with the SM by about 3 standard deviations. An authoritative overview of the experimental activities in this field can be found in [11]. A measurement of the D correlation in ^{21}Na β^{+}-decay is a major goal of the TRIμP program at KVI, Groningen [12]. FSI in this case are of the order of 10^{-4}. By measuring the angle and momentum dependence of D, new physics contributions can be distinguished from FSI down to this level. An experiment to measure the β-neutrino correlation a with ^{21}Na atoms is in progress. In order to measure D, one faces the new challenge that the radioactive sodium sample in the trap has to be polarized.

2.3 Electric Dipole Moments: Theory

A permanent electric dipole moment (EDM) of a particle, or more generally, a non-degenerate quantum system (including atoms and molecules) violates both P and T (or equivalently CP) [13]. It corresponds to a charge separation along the spin direction. The interaction Hamiltonian of a particle with spin (fixed at the origin) in a combination of magnetic and electric fields is

$$H = -\vec{\mu} \cdot \vec{B} - \vec{d} \cdot \vec{E},$$

where $\vec{\mu}$ and $\vec{d}$ are the magnetic and electric dipole moments, which, due to rotational invariance, are aligned with the spin. The experimental signal for a nonzero EDM would be the precession of the spin in a constant electric field or a linear Stark shift. In field theory, the corresponding P- and T-odd interaction between a spin-1/2 field and the electromagnetic field can be written as

$$\langle (J_{EM})_\mu \rangle = F_4(q^2) \bar{u}(p') \left[i\sigma_{\mu\nu} \gamma_5 q^\nu \right] u(p) ,$$

with $q^2 = (p' - p)^2$ and the magnitude of the EDM is given by the static limit $d = F_4(0)$ (in this notation, F_1 and F_2 would be the standard charge and magnetic moment form factors, while F_3 is a more exotic P-odd and T-even animal called the anapole). In a renormalizable field theory, a lowest-order (tree-level) coupling of this type is not allowed, but it can result from quantum loop corrections, in which case it is finite and calculable.

The search for EDMs is one of the most promising options to discover new physics [14, 15]: The EDM is an atomic physics quantity of central interest to particle physics. For many systems, high-precision experiments are possible, and the limits obtained over the years, e.g., for the neutron and electron EDM (the latter derived from measurements on mainly Cs and Tl atoms), have improved spectacularly. Moreover, as was argued above, EDM values predicted by the CKM phase in the SM are several orders below the present experimental limit. For the reasons given there, this holds for all systems: The SM value is far out of reach. On the other hand, many speculative extensions of the SM predict EDMs that are much larger than in the SM, and in many cases experimentally reachable. In such models, usually several CKM-like phases occur and, in contrast to the SM, EDMs can arise at one-loop level. A very rough, schematic estimate of the neutron in supersymmetric model would be, in units of 10^{-23} ecm,

$$d_n = K \sin\phi \left[100\,\text{GeV}/M \right]^2 ,$$

where K is some number of order 1, ϕ is some combination of CKM-like phases that parametrize CP violation, and M is the typical mass of SUSY particles. The present experimental limit on the neutron EDM shows that naive SUSY models are already getting into trouble.

An overview of the present status is given in Table 1. This table makes several important points. The most precise results have been obtained for neutral systems: ultracold neutrons, atoms, and molecules. The electron and proton EDMs were derived from EDM results for the ^{205}Tl atom and the ^{205}Tl$-$F molecule, respectively, which involve a theoretical calculation. For the charged particles muon and tau the limits are less precise. For the muon and the Λ-hyperon a limit on the EDM could be obtained from experiments in a magnetic storage ring, since the relativistic particles feel a strong electric field in their rest frames. The landscape of the EDM research field is shown in Fig. 1. On the left-hand side, the different systems that have been used to search for EDMs are shown. They vary from elementary particles to complicated atoms and molecules. Such composite systems are used for experimental reasons and because the EDMs of polarizable systems can be greatly

Table 1 EDM limits for different systems, their predictions in the SM, and typical values where nonzero signals are expected in extensions of the SM

Particle	EDM limit (95% CL)	System	Prediction SM (ecm)	New physics limit (ecm)
Electron	1.9×10^{-27}	^{205}Tl atom	10^{-38}	10^{-27}
Muon	1.05×10^{-19}	$g - 2$, rest frame $\vec{E}$	10^{-35}	10^{-22}
Tau	3.1×10^{-16}	$e^+e^- \to \tau^+\tau^-\gamma$	10^{-34}	10^{-20}
Proton	6.5×10^{-23}	^{205}Tl$-$F molecule	10^{-31}	5×10^{-26}
Neutron	7.5×10^{-26}	UCN	10^{-31}	5×10^{-26}
Λ	1.5×10^{-16}	Rest frame $\vec{E}$	10^{-30}	5×10^{-25}
^{199}Hg	2.1×10^{-28}	^{199}Hg atom	10^{-33}	5×10^{-28}

enhanced. For atoms, one distinguishes between paramagnetic atoms, the EDM of which is claimed to be due mainly to the EDM of the uncoupled valence electron, and diamagnetic atoms, where the electrons are paired and one is mainly sensitive to T-odd forces in the atomic nucleus. The shift in the ground-state energy of the mercury atom, for which the most precise EDM limit has been obtained, is of the order of $1\,\mathrm{yeV} = 10^{-24}\,\mathrm{eV}$. On the right-hand side is shown what the field is after: to find the T (or CP) violation in a fundamental theory beyond the SM at a scale of around 1 TeV. This gives rise at the scale of a few GeV to, e.g., T-odd lepton and quark EDMs and quark color-EDMs, which in turn result in nucleon EDMs and T-odd electron–nucleon and nuclear forces. When a nonzero EDM is observed, a set of complementary EDM experiments can hopefully be used to identify bottom-up the origin of the CP violation.

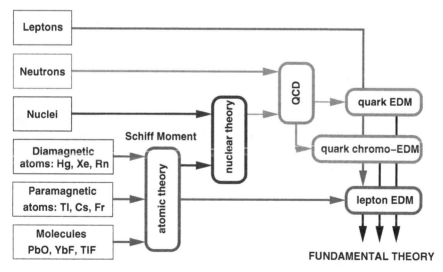

Fig. 1 The EDM landscape from experimental observations of EDMs in several systems to the underlying fundamental theory and the theoretical steps that relate the two

To relate the observations in composite systems, nucleons, nuclei, atoms, and molecules to the microscopic CP violation, theoretical calculations using hadronic and atomic theory are required. This theoretical interpretation of the EDM experiments requires a careful implementation [16] of a theorem due to Schiff [17] (for a nice review, see [18]), which in essence says that *the EDM of a nonrelativistic atom is zero*. More precisely, Schiff showed quantum-mechanically that a neutral system of charged, nonrelativistic, point-like constituents interacting only with electrostatic (Coulomb) forces will have zero EDM. This "shielding theorem" can be understood classically: The charges in a neutral composite system must rearrange themselves to screen out the effect of the external electric field.

Schiff's theorem, of course, limits severely the prospects for measuring the EDMs of electrons and nuclei through atoms. However, Schiff himself already pointed out "loopholes" provided by corrections to his theorem: Electrons in heavy atoms are relativistic; the atomic nucleus is not a point particle, but has finite size; and the electromagnetic forces in an atom are not only electrostatic, but also of magnetic origin. For diamagnetic atoms, the residual interaction for the atomic EDM involves then a new P- and T-odd moment, the so-called Schiff moment, which corresponds qualitatively to the offset of the charge and the dipole distributions in the nucleus. The calculation of the Schiff moment requires a nuclear structure model [19]. For heavy diamagnetic atoms, the net shielding factor can be estimated qualitatively as $\chi Z^2 R_{nucl}^2 / R_{atom}^2$, where Z is the atomic number, R_{nucl} is the nuclear radius, R_{atom} is the atomic radius, and $\chi \simeq 5$ is a measure of the atomic polarizability. For a system like ^{199}Hg this shielding factor is 10^{-3}–10^{-4}.

For heavy paramagnetic atoms, it was shown by Sandars in 1965 that the contribution of the electron EDM actually gets *enhanced*, in the sense that $d_{atom}/d_e \sim Z^3 \alpha^2 \chi$, where the factor $Z^2 \alpha^2$ is of relativistic origin, and the extra factor Z comes from the electric field of the nucleus: This internal electric field in atoms is much larger than any laboratory field! The factor χ is a polarizability, of order 10 for Cs. In perturbation theory, in the one-electron approximation, one can write

$$ d_{atom} = \sum_{n'} \frac{\langle ns| - d_e(\beta - 1)\, \vec{\sigma} \cdot \vec{E}\, |n'p\rangle \langle n'p| - e\, \vec{r}\, |ns\rangle}{E_{ns} - E_{n'p}} + \text{h.c.} , $$

which shows that the EDM arises from the coupling of opposite-parity s and p states and, because of the factor $\beta - 1$, it is due to the relativistic motion of the electron (β is the Dirac matrix). To estimate the enhancement factor, atomic structure calculations are required: One finds typically $d_{atom}/d_e \sim 100$, -585, 1150, $40,000$ for Cs, Tl, Fr, and Ra, respectively. After Sandars' work, ever-improving limits on the electron EDM were obtained experimentally from Cs and Tl atoms.

Even larger enhancement factors can be exploited by working with polar molecules, as suggested by Sandars in 1967. A polar molecule, such as ^{205}Tl−F, has a very large, almost fully ion-like, charge separation. As a result, the Schiff shielding is almost completely cancelled due to the fact that the effective electric field at the nucleus is as large as the externally applied electric field. For ^{205}Tl−F, the shielding over enhancement factor is about 0.67. From this diamagnetic system, a limit on the valence proton EDM has been obtained. Polar molecules, such as YbF,

PbF, and PbO, are also used to improve the limit on the electron EDM. In view of the ground-breaking developments in the cooling and trapping of cold molecules, it is likely that molecules will soon improve the best limit [20] for the electron EDM.

2.4 Electric Dipole Moments: Developments

To demonstrate the activity in the field, I have tried to give in Table 2 an overview (no doubt incomplete, and without giving references or credits) of the new genera-tion of lepton EDM experiments, using various techniques with cold paramagnetic atoms, ions, and molecules (and even solid-state experiments), with an estimate of the possible improvement over the present best limit from the ^{205}Tl atom (cf. Ta-ble 1). Many of the ideas try to exploit the huge local internal electric fields. To give an example [21], for a typical external electric field of $10\text{--}10^4$ V/cm, the valence electron in PbO feels an internal field no less than 10^{10} V/cm! For diamagnetic sys-tems, there are also interesting new developments. Research has been started on the radium atom, which has almost-degenerate opposite-parity $7s7p\,^3P_1$ and $7s6d\,^3D_2$ atomic states. A huge enhancement is predicted for the electron EDM because of this, but it remains to be seen if the degeneracy can be exploited experimentally. Also, the most neutron-rich radium isotopes have significant octupole deformation. This results in close-lying opposite-parity nuclear states, which can enhance the EDM of the nucleus by perhaps 1 or even 2 orders of magnitude. Two EDM ex-periments to measure EDM of ^{225}Ra and ^{213}Ra in atomic traps are planned, one at Argonne [22] and one at KVI [12]. Other experiments on diamagnetic atoms are for liquid ^{129}Xe and ^{223}Rn; also the ^{199}Hg experiment [23] can still be improved significantly. Table 3 contains (with the same caveats as for Table 2) an overview of the new generation of EDM for hadronic systems. Several new neutron EDM [24] experiments are underway or planned, using more intense UCN sources or other improvements, and aiming for $10^{-27}\text{--}10^{-28}$ ecm.

Table 2 The new generation of leptonic EDM searches

Group	System	Advantages	Improvement
D. Weiss (Penn State)	Cs optical lattice	Long coherence	400
D. Heinzen (Texas)	Trapped Cs	Long coherence	100?
H. Gould (LBL)	Cs fountain	Long coherence	100?
L. Hunter (Amherst)	GdIG solid	Huge S/N	100?
S. Lamoreaux (LANL),			
C.-Y. Liu (Indiana)	GGG solid	Huge S/N	$100\text{--}10^5$?
E. Hinds (Imperial)	YbF beam/trap	Int. $\vec{E}$, long T	$10\text{--}100$
D. DeMille (Yale)	PbO* cell	Int. $\vec{E}$, good S/N	$2\text{--}100$?
E. Cornell (JILA)	Trapped HfH$^+$ ion	Int. $\vec{E}$, huge T	100?
N. Shafer-Ray (Okla.)	PbF beam/trap	Int. $\vec{E}$, long T	100?
L. Willmann (KVI)	^{213}Ra	Atomic enhancement	10^4?
J. Miller, Y. Semertzidis,			
Y. Kuno (J-PARC)	Muon	Storage ring	$10^5\text{--}10^6$

Table 3 The new generation of hadronic EDM searches

Group	System	Advantages	Improvement
D. Wark, M.v.d. Grinten			
(Sussex/RAL, ILL)	UCN	Cryogenic	10–100?
S. Paul (Münich)	UCN		?
O. Naviliat-Cuncic, K. Kirch (PSI)	UCN	Neutron intensity	10–100?
S. Lamoreaux, M. Cooper,	UCN in		
J.C. Peng (LANSCE, SNS)	superfluid ^{4}He	^{3}He comagnetometer	100–10^3?
N. Fortson (Washington)	^{199}Hg vapor cell		3?
M. Romalis (Princeton)	Liquid ^{129}Xe	Density, long T	100–10^5
T.E. Chupp, C.E. Svensson	^{223}Rn cell	Nucl. enhanc.	10–100?
(TRIUMF)			
Z.-T. Lu, R. Holt (ANL)	Trapped ^{225}Ra	At. + nucl. enhanc.	?
L. Willmann (KVI)	^{213}Ra, ^{225}Ra	At. + nucl. enhanc.	?
J. Miller, Y. Semertzidis,			
E. Stephenson (BNL, CERN?)	Deuteron	Storage ring	10–10^3

A novel idea is to use dedicated magnetic storage rings to measure the EDM of specific *charged* particles [25, 26]. Relativistic particles in such a ring feel a very strong motional electric field in their rest frame, resulting in a spin precession out of the orbital plane. An additional radial electric field in the ring can be used to compensate the spin precession (in the orbit) due to the magnetic moment anomaly, provided the effective anomaly is small. An experiment is planned for muons at J-PARC with a sensitivity of about 10^{-24} ecm, cf. Table 2. Also the deuteron is suitable for this approach. An experiment is planned that targets a sensitivity of 10^{-27} ecm. This is very interesting since it has been shown [27] that the sensitivity of the deuteron EDM to CP-violation scenarios, which comes predominantly from the T-odd two-body nuclear force [28], is complementary to that of the neutron. Moreover, in certain models the deuteron is significantly more sensitive than the neutron. For example, in specific cases where the EDM comes mainly from quark color-EDMs, the deuteron and neutron EDMs are given by

$$d_D = -4.67 d_d^c + 5.22 d_u^c \,,$$
$$d_n = -0.02 d_d^c + 0.49 d_u^c \,,$$

in terms of the color-EDMs of the up- and down-quark.

The key question for these hadronic systems is how they compare in terms of sensitivity to the underlying CP violation. In Table 4, I have collected the results of pertinent calculations for some of the most promising cases of Table 3 (in the case of ^{129}Xe, ^{199}Hg, and ^{225}Ra, of course, complicated nuclear many-body calculations are involved, with large theoretical uncertainties). The table lists the short-term and final goal of the planned experiments. The results of the calculations are expressed in terms of three independent P- and T-odd pion–nucleon coupling constants that parametrize the microscopic CP violation. These columns should be added: e.g., the neutron EDM is given by $d_n = 0.14(g_0 - g_2)$ and the deuteron EDM is $d_D = 0.10 g_0 + 0.23 g_1$ (the strategy and the details of these calculations can be

Table 4 Sensitivity to CP violation

	Short-term goal (ecm)	Final goal (ecm)	g_0	g_1	g_1	Limit on θ
n	5×10^{-27}	5×10^{-28}	0.14		-0.14	2×10^{-12}
D	10^{-27}	10^{-29}	0.10	0.23	0.00	10^{-13}
^{129}Xe	10^{-30}–10^{-31}	10^{-33}	6×10^{-5}	6×10^{-5}	12×10^{-5}	10^{-13}
^{199}Hg	5×10^{-29}	–	2×10^{-6}	2×10^{-4}	-3×10^{-5}	5×10^{-10}
^{225}Ra	?	?	-0.06	-0.12	0.11	?

found in [27]). It is natural to assume that g_0, g_1, and g_2 are of the same order in specific models, so the table provides a model-independent estimate of the sensitivity of the different cases. It shows that n and D are complementary and of similar sensitivity; ^{129}Xe is intrinsically some 10 times less sensitive than ^{199}Hg; the deuteron and ^{129}Xe experiments at their final goals are comparable. The enhancements in the radium atom (for which no experimental goals are stated yet) are so large that they overcome the Schiff screening. In the last column the limit on the QCD θ angle that can be achieved is given, in the special case that it is θ that gives rise to the EDMs.

3 Is the Neutrino Its Own Antiparticle?

3.1 Neutrino Oscillations and Their Consequences

In the last decade neutrino physics has gone through spectacular developments (for the status, see [29]). In the years 2002–2004, neutrino oscillations became established after experiments with atmospheric [30], solar [31, 32], reactor [33, 34], and accelerator [35] neutrinos. It is now beyond doubt that neutrino flavors mix and that neutrinos have mass. The solar neutrino puzzle [36, 37] is solved: The missing solar neutrinos in Davis' Homestake experiment are explained by the fact that two-thirds of the solar ν_e-flux on its way to earth changed into a flavor that the experiment could not detect. The experiment was correct, and the fluxes of ^{8}B neutrinos determined by SNO's charged-current, elastic scattering, and neutral-current results and the Super-Kamiokande flux are in good agreement with the predictions of Bahcall's Standard Solar Model. We understand well how the sun shines [38].

In the SM of particle physics, neutrinos come in three flavors, ν_e, ν_μ, and ν_τ; it was assumed that neutrinos were massless and purely left-handed [39]. We now know that neutrinos have three mass eigenstates ν_i ($i = 1, 2, 3$), which are different from the weak interaction, flavor, eigenstates. Lepton family number is now proven to be violated, although total lepton number is still conserved. In a weak decay, a neutrino is produced in a flavor eigenstate, ν_α, which is a quantum-mechanical superposition of the mass eigenstates, ν_i:

$$|\nu_\alpha\rangle = \sum_i U_{\alpha i}^* |\nu_i\rangle \, ,$$

where U is a 3×3 unitary mixing matrix. After propagation over a distance L, this state has evolved into (we assume the neutrino mass is very small)

$$|\nu_\alpha(L)\rangle = \sum_i U_{\alpha i}^* e^{-i(m_i^2/2E)L}|\nu_i\rangle ,$$

from which follows

$$|\nu_\alpha(L)\rangle = \sum_\beta \left[\sum_i U_{\alpha i}^* e^{-i(m_i^2/2E)L} U_{\beta i} \right] |\nu_\beta\rangle ,$$

i.e., ν_α has turned into a superposition of *all* flavors.

The unitary mixing matrix U, the analogon of the CKM matrix for the quarks, called the Pontecorvo-Maki-Nakagawa-Sato (PMNS) matrix, can be parametrized as

$$U = \begin{pmatrix} 1 & 0 & 0 \\ 0 & c_{23} & s_{23} \\ 0 & -s_{23} & c_{23} \end{pmatrix} \begin{pmatrix} c_{13} & 0 & s_{13}e^{-i\delta} \\ 0 & 1 & 0 \\ -s_{13}e^{i\delta} & 0 & c_{13} \end{pmatrix} \begin{pmatrix} c_{12} & s_{12} & 0 \\ -s_{12} & c_{12} & 0 \\ 0 & 0 & 1 \end{pmatrix} \begin{pmatrix} e^{i\alpha} & 0 & 0 \\ 0 & e^{i\beta} & 0 \\ 0 & 0 & 1 \end{pmatrix} ,$$

where $c_{ij} = \cos\theta_{ij}$ and $s_{ij} = \sin\theta_{ij}$. The product of the first three matrices is just like the quark case: It contains three mixing angles and one phase, δ, that gives rise to CP violation in the lepton sector. The first matrix describes atmospheric neutrino oscillations, while the third matrix describes solar neutrino oscillations. The corresponding mixing angles $\theta_{23} = 45°(6)$ and $\theta_{12} = 32.5°(2.5)$ have been determined quite accurately; surprisingly, they are large. The mixing angle in the second matrix has not been determined yet, but an upper limit $\theta_{13} < 14°$ was obtained by the Chooz reactor neutrino-oscillation experiment. The Double-Chooz experiment, which will start in 2007, aims to measure θ_{13}. If it turns out that θ_{13} is not too small, there is hope that future CP-violation experiments will be able to measure the phase δ, as one can understand from the structure of the second matrix.

The two phases α and β in the fourth matrix are present if neutrino are Majorana fermions. If neutrinos are Dirac fermions, these phases can be removed by redefining the neutrino fields. Nobody has a clue how they could realistically be measured [40]. This brings us to what is the most fundamental fact which, despite all the recent progress in neutrino physics, we do not know (yet) about neutrinos: Are they Dirac or Majorana fermions? If the neutrino is a Dirac fermion, the antineutrino is a distinct particle, but if the neutrino is a Majorana fermion, then it cannot be distinguished from the antineutrino. Because neutrinos are neutral [41], they are unique in allowing both options, while the situation is clear for the charged leptons, because they carry the additive quantum number charge.

Why, you may ask, is it not known yet whether the neutrino is a Dirac or a Majorana fermion? The answer [42] is historically relevant and conceptually instructive. Let us define the neutrino as the particle that creates electrons, while the antineutrino is the particle that creates positrons, e.g.,

$$\nu_e + n \rightarrow p + e^- \,,$$
$$\overline{\nu}_e + p \rightarrow n + e^+ \,.$$

Now if $\nu_e \equiv \overline{\nu}_e$, the antineutrino produced in the decay $n \rightarrow p + e^- + \overline{\nu}_e$ can create an electron by converting on a neutron. Unfortunately, this is hard to test experimentally, since two weak interactions are involved. However, in a nucleus, a second-order weak decay, two-neutrino double-β decay ($2\nu\beta\beta$),

$$2n \rightarrow 2p + 2e^- + 2\overline{\nu}_e \,,$$

can sometimes occur, when ordinary β-decay is energetically forbidden. Bu then, if $\nu_e \equiv \overline{\nu}_e$, the decay mode *without* neutrinos

$$2n \rightarrow n + p + e^- + \overline{\nu}_e \equiv n + p + e^- + \nu_e \rightarrow 2p + 2e^-$$

should *also* occur; in fact, since it is favored by a larger phase space, it should occur more often. In the early 1950s, the absence of experimental evidence for this *neutrinoless double-β decay* ($0\nu\beta\beta$) process was interpreted as proof that the neutrino was a Dirac fermion. The concept of *lepton number* was introduced, where the value $+1$ was assigned to the electron and the neutrino and the value -1 to the positron and the antineutrino. All the empirical findings then simply follow from lepton-number conservation. In particular, the $2\nu\beta\beta$ process is allowed, while $0\nu\beta\beta$ is forbidden, since it violates lepton number by two units.

However, after maximal parity violation had been discovered in 1957, it became clear that this reasoning was flawed, since the massless Majorana neutrino in

$$2n \rightarrow n + p + e^- + \overline{\nu}_e^{RH} \equiv n + p + e^- + \nu_e^{RH}$$

is right-handed, and thus it has the wrong helicity to convert on the neutron, which requires a left-handed neutrino, viz.,

$$\nu_e^{LH} + n \rightarrow p + e^- \,.$$

Therefore, the nonobservation of neutrinoless double-β decay does *not* imply that neutrinos are Dirac fermions or that lepton number is conserved. It could be that $0\nu\beta\beta$ is simply suppressed by the lightness of the Majorana neutrino and the corresponding helicity mismatch. In other words, instead of concluding that neutrinos are Dirac fermions, all experimental results can alternatively be explained by replacing

$$\nu_e \rightarrow \nu_e^{LH} \,,$$
$$\overline{\nu}_e \rightarrow \nu_e^{RH} \,,$$

together with maximal parity violation. Any difference between the two scenarios, Dirac or Majorana, is proportional to the very small number $(m_v/E_v)^2$, where $m_v \simeq$ 50 meV, say, and a typical E_v is of the order of MeVs or tens of MeV.

3.2 The Absolute Neutrino Mass

Apart from the mixing angles and the phases, the neutrino mass matrix contains of course the three masses m_i ($i = 1, 2, 3$). Oscillation experiments, however, are only sensitive to $\Delta m_{ij}^2 = m_i^2 - m_j^2$, the differences of the squares of the neutrino masses, but they provide no direct information on the individual neutrino masses m_i themselves. The atmospheric and solar experiments have measured $|\Delta m_{23}^2| =$ $2.5(6) \times 10^{-3}\,\mathrm{eV}^2$ and $\Delta m_{12}^2 = 8.2(6) \times 10^{-5}\,\mathrm{eV}^2$. Oscillation experiments can determine only six out of the nine parameters of the neutrino mass matrix; of these, four have been determined (one up to its sign), and an upper limit on one has been obtained. This leaves us with several options for the neutrino-mass spectrum, cf. Fig. 2. The spectrum can have a *normal* hierarchy (like the quarks) or an *inverted* hierarchy. From the absolute measurement of one neutrino mass, the whole spectrum could be constructed. The most natural guess for the absolute mass scale is $\sqrt{|\Delta m_{23}^2|} \simeq 0.05\,\mathrm{eV}$. However, it is still possible that it is larger, of the order of 0.1–1 eV. In that case, the spectrum is called *degenerate*.

It is crucial both for particle physics and for astrophysics and cosmology to determine the absolute neutrino mass scale. But how to go about it? There are several options:

(a) The established way is to determine the neutrino mass directly from the kinematics of a nuclear β-decay, traditionally tritium because it has a low Q-value

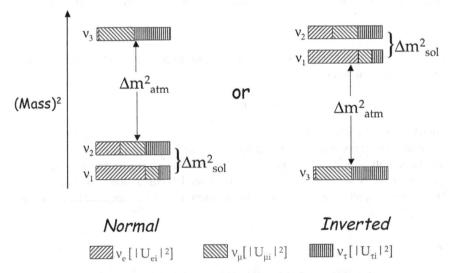

Fig. 2 The neutrino mass spectrum; *left*: normal hierarchy, *right*: inverted hierarchy

of 18.6 keV. This was already suggested by Fermi himself. The end-point of the electron spectrum in tritium decay is the most sensitive part of the phase space, because the electrons then take almost all the available energy, while the neutrino becomes nonrelativistic. Ignoring neutrino mixing, the energy spectrum of an allowed β-decay is given by

$$\frac{dN}{dE} = f(E)(E_0 - E_e)\sqrt{(E_e - E_0)^2 - m_v^2}\,\theta(E_0 - E_e - m_v)\,,$$

where E_0 is the total decay energy, E_e is the electron energy, and the step-function θ reflects that the mass m_v can only be produced when there is sufficient energy available. The dependence on m_v comes only from phase-space factors, the function $f(E) = G_F^2 \cos^2 \theta_C \, m_e^5/(2\pi^3)|M_{\mathrm{nucl}}|^2 F(Z,E)pE$ does not depend on m_v. Each neutrino mass eigenstate m_i would result in a kink in the spectrum at $E_0 - m_i$; however, because of their limited energy resolution, tritium β-decay experiments cannot, in fact, see these kinks. They are sensitive to an average "electron-neutrino mass" defined by

$$m^2(v_e) \equiv \sum_i |U_{ei}|^2 m_i^2\,,$$

cf. the discussion in [43]. The best limit on the (anti)neutrino mass from tritium β-decay at present stands at about 2 eV, e.g., the final result from the Mainz experiment is $m_{\bar{v}_e} < 2.3$ eV (95% CL) [44].

(b) Neutrinoless double-β decay, which is discussed in more detail in Sect. 3.3, is another option to obtain information on the absolute neutrino mass. The transition matrix for the decay process is directly proportional to the effective neutrino mass:

$$m_{ee} = \sum_i |U_{ei}^2 m_i|\,,$$

which is the coherent sum over the neutrino mass eigenstates that contribute to the electron-neutrino flavor eigenstate with mixing-matrix elements U_{ei}. The problems here are the unknown Majorana phases that enter m_{ee} and the large uncertainties in the nuclear structure part of the decay matrix element; cf. the discussion below.

(c) The neutrino mass can nowadays also be limited indirectly from cosmology. Relic neutrinos from the Big Bang are thought to contribute to "hot" (relativistic particles) dark matter. Analyses of the fluctuations in the cosmic microwave background measured by the WMAP satellite, together with analyses of the large-scale structure of the distribution of galaxies and information from the Lyman-α forest or X-ray clusters, are now so precise that a competitive value for the neutrino mass comes out. If the mass is too high, neutrinos become nonrelativistic too soon, creating too much small-scale substructure, while if it is too small, they stay relativistic too long, wiping out structure at small scales. This approach is sensitive to the total mass of the neutrinos, $\sum_i m_i$. Upper limits reported are in the 0.5 eV range [45]. The analyses, however, are not entirely model independent.

(d) Another, direct, astrophysical way to limit the neutrino mass is from the time of flight of neutrinos emitted in supernovae. If neutrinos have mass, their velocity is no longer the speed of light, $c = 1$, but instead

$$v = \frac{p_\nu}{E_\nu} \simeq 1 - \frac{m_\nu^2}{2E_\nu^2} .$$

The difference in their arrival time on earth then depends on their energy as

$$\delta t/t = \delta v/v \simeq \frac{m_\nu^2}{E_\nu^2} \frac{\delta E_\nu}{E_\nu} .$$

On February 23, 1987, some 20 neutrinos from supernova SN1987A in the Large Magellanic Cloud (about 50 kpc away, so $t \simeq 5 \times 10^{12}$ s) were detected within a time span of 10 s, with energies around 20–30 MeV and an energy spread of about 10 MeV. The analysis of these few events resulted in an upper limit on $m_{\bar{\nu}_e}$ of 10–20 eV [46, 47]. Imagine the amount of money that physicists had spent to arrive at a similar result! On the other hand, because of the large uncertainties in existing models of supernova explosions, it will probably be impossible to improve this method to sub-eV neutrino masses, in the case of future supernovae.

There is no substitute for a direct laboratory determination of the neutrino mass that is purely kinematic, such as tritium β-decay, as long as this method has not run out of steam. Since, for $m_\nu = 1$ eV, a fraction of only 2×10^{-13} of all tritium decays occurs in the last eV below the end-point, it is clear that such experiments are very demanding, requiring high-energy resolution, large acceptance, low background, and of course lots of tritium. The KATRIN experiment [48], which is presently being set up at Karlsruhe, has the ambitious goal to improve the limit on $m_{\bar{\nu}_e}$ by an order of magnitude. It will have a sensitivity of 0.2 eV, where a nonzero neutrino mass of 0.3 eV would be measured with 3 σ significance and a mass of 0.35 eV with 5 σ. This sensitivity of KATRIN is cosmologically relevant, it is in the region where a nonzero $0\nu\beta\beta$ signal has been claimed (see below), and it will be able to distinguish between degenerate and hierarchical neutrino mass spectra.

3.3 Neutrinoless β-Decay

The only practical experiment to settle the question about the nature of the neutrino (Dirac or Majorana) is *neutrinoless double-β decay* (for reviews, see, e.g., [42, 49]). This is a nuclear decay that changes the charge by two units. The nuclear pairing energy makes certain even–even nuclei are relatively stable compared to odd–odd ones that they would decay by ordinary β-decay. In selected cases, the two-neutrino double-β decay process

$$(Z, A) \rightarrow (Z+2, A) + 2e^- + 2\bar{\nu}_e$$

can take place, while the single β-decay process

$$(Z,A) \to (Z+1,A) + e^- + \overline{v}_e$$

is energetically forbidden (or heavily suppressed). The 2νββ process is a second-order weak interaction that is allowed in the SM, but the lifetime is very long; in fact, 2νββ is the rarest process ever observed in Nature. Both for 0νββ and 2νββ the quantity measured is the half-life, which for 2νββ is beyond 10^{20} years. The experimental signal is the coincidence of two electrons, the energy of which adds up to the Q-value of the decay. The sum energy spectrum is continuous for 2νββ, while it would be a peak right at the Q-value for 0νββ.

In Table 5 the half-life limits obtained for the different isotopes used for 0νββ experiments are listed. These isotopes were chosen because they have a relatively high Q-value; the 0νββ decay rate scales as Q^5 and the 2νββ decay rate as Q^{11}. Also, the natural abundance of these isotopes is reasonable and, in some cases, can be increased by using enriched samples. Currently the best 0νββ limit comes from experiments with ^{76}Ge, in particular the Heidelberg–Moscow experiment in the Gran Sasso underground laboratory [50], which probe lifetimes longer than 10^{25} years (i.e., corresponding to one decay event per kg per year!). This generation of experiments uses of the order of 10 kg of active material. The next generation of 0νββ experiments, cf. Table 6 [51], will use tons of material and will target lifetimes well beyond 10^{27} years, or effective neutrino masses of 10–50 meV! Part of the Heidelberg–Moscow collaboration has claimed [52], based on additional data and an alternative statistical analysis, that their results imply the detection of 0νββ decay, instead of being just an upper limit, cf. Table 5. It corresponds to a best value for the effective neutrino mass of about 0.4 eV. This claim is regarded as controversial by the community, but it will soon be confirmed or ruled out by other 0νββ experiments or by the KATRIN tritium β-decay experiment.

The first 0νββ observation will be a landmark event: We can conclude immediately that neutrinos are Majorana fermions [53] and that lepton number is violated. However, a full interpretation of the experiments, and in particular the extraction

Table 5 Limits obtained for the different isotopes for 0νββ decay

Parent	Daughter	Half-life (year)	CL (%)	$\langle m_v \rangle$ (eV)
^{48}Ca	^{48}Ti	$> 9.5 \times 10^{21}$	75	< 8.3
^{76}Ge	^{76}Se	$> 1.9 \times 10^{25}$	90	< 0.35
^{76}Ge	^{76}Se	$0.7–4.2 \times 10^{25}$	90	$0.2–0.6$
^{82}Se	^{82}Kr	$> 2.7 \times 10^{23}$	90	< 5.0
^{100}Mo	^{100}Ru	$> 5.6 \times 10^{23}$	90	$< 0.6–2.0$
^{116}Cd	^{116}Sn	$> 1.7 \times 10^{23}$	90	< 1.7
^{128}Te	^{128}Xe	$> 7.7 \times 10^{24}$	68	< 1.1
^{130}Te	^{130}Xe	$> 2.2 \times 10^{24}$	90	$< 0.2–1.1$
^{136}Xe	^{136}Ba	$> 4.4 \times 10^{23}$	90	< 2.3
^{150}Nd	^{150}Sm	$> 2.1 \times 10^{21}$	90	< 4.1

Table 6 The new generation of $0\nu\beta\beta$ experiments [51]

Experiment	Isotope	Approach
CANDLES	^{48}Ca	Several tons of CaF$_2$ crystals in liquid scintillator
CARVEL	^{48}Ca	100 kg ^{48}CaWO$_4$ crystal scintillators
COBRA	^{116}Cd, ^{130}Te	420 kg CdZnTe semiconductors
CUORE	^{130}Te	750 kg TeO$_2$ cryogenic bolometers
DCBA	^{150}Nd	20 kg Nd layers between tracking chambers
EXO	^{136}Xe	1 ton Xe TPC (gas or liquid)
GERDA	^{76}Ge	40 kg Ge diodes in LN$_2$, expand to larger masses
GSO	^{160}Gd	2 ton Gd$_2$SiO$_3$, Ce crystal scintillator in liquid scintillator
MAJORANA	^{76}Ge	180 kg Ge diodes, expand to larger masses
MOON	^{100}Mo	Several tons of Mo sheets between scintillator
SNO++	^{150}Nd	1000 ton Nd-loaded liquid scintillator
SuperNEMO	^{82}Se	100 kg Se foils between TPCs
Xe	^{136}Xe	1.56 ton Xe in liquid scintillator
XMASS	^{136}Xe	10 ton liquid Xe

of the effective neutrino mass, is hampered by the poor knowledge of the nuclear matrix element that occurs in the transition amplitude of the process. The effective neutrino mass in $0\nu\beta\beta$ can be determined from the half-life $\tau_{1/2}$ by

$$\left[\tau_{1/2}\right]^{-1} = G_{\text{phs}}|M_{\text{nucl}}|^2 \frac{\langle m_{\text{ee}}\rangle^2}{m_{\text{e}}^2} \, ,$$

where G_{phs} is the phase-space factor and $|M_{\text{nucl}}|^2$ is the nuclear matrix element; the expression for m_{ee} in terms of the PMNS parameters was given above. To translate the measured lifetime into a Majorana neutrino mass, the value of the nuclear matrix element is needed at some level of accuracy. How well can nuclear theory calculate such complicated many-body quantities? A valid, but pessimistic point of view [54] is to realize that it will not be possible in the near future to derive nuclear matrix elements for large A from QCD. This means that one has to live with the published range of calculated matrix elements, which for the important case of ^{76}Ge, for example, covers some 2 orders of magnitude [54]. Within a particular many-body approach the spread is, of course, significantly less, but the calculations may still not be close to the true value.

Nuclear theory could be guided by certain measurements involving the relevant isotopes for $0\nu\beta\beta$, as input to constrain their calculations. The allowed $2\nu\beta\beta$ process, which has been observed in 12 isotopes, is of obvious relevance here. However, one has to realize that the typical excitation energies (a few MeV) are significantly lower than for the $0\nu\beta\beta$ process (of order 10 MeV). Moreover, in $2\nu\beta\beta$ the coupling is only to virtual 1^+-states in the intermediate nucleus, while in $0\nu\beta\beta$ intermediate states of all multipoles contribute. Important experiments are also charge-exchange reactions [55, 56] that measure the transition strength to the virtual states in the intermediate nucleus. Pertinent measurements of $(d, ^2\text{He})$ and $(^3\text{He}, t)$ reactions have been performed at KVI and RCNP, respectively, e.g., for the $0\nu\beta\beta$-triple

^{48}Ca–^{48}Sc–^{48}Ti, and more could be done at facilities like KVI, PSI, and RCNP. At the same time, it is crystal clear that new theoretical ideas are direly needed as well.

3.4 The Seesaw: Another Playground for Theorists

Let us finally discuss some implications of the neutrino nature and its mass for particle physics. As we now know that neutrinos do have mass, we have to address the question how to extend the SM to account for this experimental fact. Many theorists favor one particular candidate scenario: the *seesaw model*. This is the minimal extension of the SM that adds right-handed neutrinos and includes all renormalizable interaction allowed by the $SU(3) \times SU(2) \times U(1)$ gauge symmetry. Right-handed neutrinos have no interactions with the SM gauge bosons (they are "sterile"), they can only have Yukawa couplings to the Higgs scalar and the lepton doublets. The celebrated *seesaw mechanism* then provides a possible answer to the otherwise mysterious fact that neutrinos are so much lighter than the other SM fermions, the quarks and the charged leptons. In its simplest form, the seesaw mechanism predicts generically that

$$m_\nu = \frac{m_D^2}{m_R},$$

where m_D is a Dirac mass, typically equal to the mass of quarks or charged leptons (which get their mass in the SM from the Higgs mechanism), and m_R is a right-handed Majorana neutrino mass. Since right-handed neutrinos have no interactions in the SM, they can have Majorana mass terms, with a mass that is independent of electroweak symmetry breaking, and thus can be very high. If we take m_D of the order of the electroweak scale (say 200 GeV), neutrino masses of the order of 50 meV, as suggested by the oscillation data, go together with an m_R of the order of 10^{14} GeV. These values show also why this is called the seesaw mechanism: Neutrinos are very light because they are connected to right-handed neutrinos that are very heavy, where the rotation point of the seesaw is the electroweak scale.

Theorists also get a kick out of the seesaw model because it is a renormalizable field theory, which implies that it has a finite number of parameters, and it is natural in the sense that *all* interactions allowed by the SM gauge symmetries are included in the model. It also fits nicely with the theoretical idea of grand unification, since the seesaw scale m_R suggested by the neutrino-oscillation experiments is consistent with the energy scale at which the unification of the forces is supposed to occur. The seesaw mechanism, in fact, can be easily embedded in many GUTs. Finally, the seesaw model offers an elegant and rather compelling solution [57, 58] to the cosmological dominance of matter over antimatter: The heavy right-handed neutrinos undergo CP-violating decays in the early universe, thus generating an asymmetry of leptons of antileptons ("leptogenesis") which subsequently can result in baryogenesis.

Of course, one can also assume that neutrinos are Dirac fermions and simply add the right-handed neutrinos to the SM Lagrangian. Neutrinos then get Dirac masses upon electroweak symmetry breaking by Yukawa interactions with the Higgs field, in the same way as all the other SM fermions. In this scenario, lepton number is conserved, so there is no leptogenesis, and the smallness of the neutrino masses is then just accidental. For these reasons, the seesaw model with Majorana masses would appear to be preferred. At low energy, it gives rise to a Majorana mass term for the left-handed neutrinos, implying a PMNS-type mixing matrix and lepton-number violation [59]. Other terms are suppressed by additional inverse powers of the scale m_R [60, 61], and for that reason, unfortunately, it is very hard to test experimentally. Another very clever idea in neutrino physics is needed here.

4 Atomic Parity Violation and the Weinberg Angle

4.1 The Unified Electroweak Theory

In the electroweak part of the SM, the weak interactions of quarks and leptons are unified with QED, the quantum field theory describing the electromagnetic interaction mediated by massless photons. This unified theory claims that all electromagnetic and weak phenomena are manifestations of one universal electroweak interaction among spin-1/2 quarks and leptons, mediated by four massless spin-1 bosons, two charged and two neutral ones. The short range of the weak interaction is due to the large masses that three of the four bosons acquire from the spontaneous breaking of the electroweak symmetry by the vacuum, possibly resulting from a phase transition in the early universe.

In the electroweak theory, the two neutral bosons mix, where one linear combination, the photon, γ, remains massless and couples to electric charge. The existence of the second neutral boson, the Z^0, with a mass of about 91 GeV, is required for the consistency of the theoretical framework. A whole new class of weak phenomena, the so-called neutral-current interactions, mediated by the Z^0, was predicted by the theory. The mixing of the photon and the Z^0 is described by one fundamental parameter, the *weak mixing angle* θ_w, frequently called the Weinberg angle. This mixing angle connects the two independent coupling constants of the electroweak theory: the electric charge, e, and the coupling constant of the weak interaction, g_w, via the relation $\sin^2 \theta_w = e^2/g_w^2$. Neutral currents were discovered in the early 1970s and the $W^\pm$ and Z^0 bosons in the early 1980s in high-energy scattering processes at CERN. Since then, accelerator experiments have confirmed the predictions of the electroweak theory at high energies with great precision.

Because the electroweak theory is a quantum field theory, the coupling "constants" are actually not constant, but they vary with the four-momentum scale at which they are measured (or, alternatively, with the distance at which the interaction is probed). This variation, or *running*, with four-momentum scale, Q, is a quantum effect: It is caused by the creation and annihilation of particle–antiparticle pairs in the vacuum. These clouds of virtual particles shield the interacting particles by polarizing the vacuum. This results in an effective, scale-dependent, γ-Z^0 mixing angle which is related to the running coupling constants through

$$\sin^2 \theta_w(Q) = e^2(Q)/g_w^2(Q) \,.$$

In Fig. 3 the prediction of the electroweak theory for the running of the mixing angle is plotted. It demonstrates that $\sin^2 \theta_w$ first decreases, by some 3%, by going from low to high scales. This is due to vacuum polarization by quark–antiquark pairs, causing g_w to increase. Unlike photons, however, $W^\pm$ and Z^0 bosons couple to each other, and above about 100 GeV, $W^\pm$ pairs start to dominate the vacuum polarization, leading to "antishielding" and a decrease of g_w. The latter is similar to the running of the coupling constant of the strong interaction between quarks due to vacuum polarization by gluon pairs, resulting in the vanishing of the coupling at very high scale, or zero distance, the famous "asymptotic freedom" of quarks.

4.2 Atomic Parity Violation

Because the Z^0 is neutral, it can interfere with photon exchange, but since it is so heavy the effects of Z^0 exchange are in general tiny compared to the electromagnetic force. However, one of the most striking properties of the weak interaction is that, unlike the electromagnetic interaction, it is not left–right symmetric, it violates parity: The $W^\pm$ bosons couple only to *left-handed* (clockwise spinning) fermions, and also the interactions of the Z^0 violate parity. This provides a very clean filter to detect experimentally the effects of Z^0 exchange. In an atom or ion, Z^0 exchange manifests itself in a very small breakdown of standard parity selection rules: atomic parity violation [63] (for reviews, see [64, 65]).

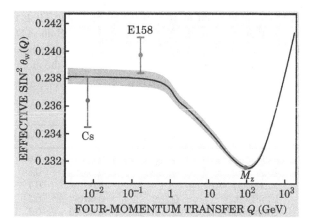

Fig. 3 The "running" of the γ-Z^0 mixing angle from low to high momentum scale [62]. The *curve* is the prediction of the SM. The experimental points are explained in the text

Consider the electroweak interaction between an atomic electron and the quarks in a nucleus with Z protons and N neutrons, so with $2Z + N$ up-quarks and $Z + 2N$ down-quarks. The dominant contribution to atomic parity violation comes from the exchange of the Z^0 between the electron and the quarks, where the Z^0 couples to the spin of the electron and to the weak charge of the quarks. Then, just like for the electric charge, all the weak charges of the quarks add up, resulting in the total *weak charge*, Q_w, defined by

$$Q_w = (2Z + N)Q_w(u) + (Z + 2N)Q_w(d)$$
$$= -N + (1 - 4\sin^2\theta_w)Z ,$$

where the second line follows from the values of the weak charges of the u- and d-quarks according to the electroweak theory. This weak charge Q_w is the experimental observable from which the weak mixing angle can be deduced. It plays the same role for the electroweak interaction due to Z^0 exchange as the electric charge does for the Coulomb interaction, and it arises from the coherent effect of all the quarks in the atom or ion. Corrections to this dominant Z^0-exchange contribution to Q_w come from one-loop vacuum-polarization diagrams discussed above; their size can be predicted accurately by the electroweak theory. Thus, from a measurement of Q_w the value of the effective mixing angle at low momenta can be extracted and compared with the prediction from Fig. 3.

The running of $\sin^2\theta_w$ is still a poorly tested prediction of the quantum structure of the electroweak theory. As can be seen from Fig. 3, the mixing angle has been accurately measured at high Q by scattering experiments at high energies, comparable to the mass of the Z^0. A recent parity-violating electron–electron scattering experiment by the SLAC E158 collaboration [66, 67] has resulted in a value for the mixing angle at intermediate Q that is in reasonable agreement with the SM. However, a recent neutrino scattering experiment by the NuTeV collaboration [68] at comparable Q disagrees with the prediction of the SM. At Jefferson Laboratory, the Qweak experiment is planned, which aims to determine the mixing angle by measuring parity violation in electron–proton scattering. Table 7 lists the recent and planned experiments that measure the Weinberg angle. E158 and Qweak measure the weak charges of the electron and proton, respectively, at intermediate Q, whereas in an atomic parity-violation experiment the weak charge of the atom (or ion), i.e.,

Table 7 Experiments that determine the Weinberg mixing angle

Measurement	Laboratory	Observable	$\delta\sin^2\theta_w$	$\delta\sin^2\theta_w / \sin^2\theta_w$ (%)
e^+e^- at Z^0-pole	LEP@CERN		0.00017	0.07
Cs APV	JILA Boulder	$Q_w(\text{Cs})$	0.0016	0.6
ν-DIS	NuTeV@FNAL		0.0016	0.7
PV ee scattering	E158@SLA	$Q_w(e)$	0.0013	0.5
PV ep scattering	Qweak@TJNAF	$Q_w(p)$	0.00072	0.3
PV ee scattering	TJNAF	$Q_w(e)$	0.00025	0.1
PV eD DIS	TJNAF		0.0011	0.45

the weak charges of all the quarks acting coherently, is measured at low Q. The momentum scale involved in atomic parity violation is around $Q \simeq 1\,\text{MeV}/c$, while for the high-energy accelerator experiments $Q \simeq 100\,\text{GeV}/c$. Hence, an atomic parity-violation experiment from which a sufficiently precise value for $\sin^2\theta_{\rm w}$ could be extracted would, together with the experiments at intermediate energy, map out the running of the γ-Z^0 mixing angle over some 5 orders of magnitude in momentum scale.

The field of atomic parity violation emerged in 1974 [63], right after the discovery of neutral currents. Candidates for actual experiments are heavy atoms or ions with one valence electron, since the weak charge scales with Z^3, and, moreover, for such systems the atomic theory can be pushed to great (sub-1%) accuracy. Examples are the cesium and francium atoms and the barium and radium ions; see below. The benchmark is set at present by the high-precision measurement performed by the Boulder group of Weiman and collaborators in an atomic beam experiment [69, 70]. They measured the weak charge of cesium with a precision of about 0.4%. The value for the mixing angle extracted from their experimental value for the cesium weak charge is also plotted in Fig. 3; it is not precise enough to confirm the predicted running of $\sin^2\theta_{\rm w}$; it is, in fact, also still consistent, within about two standard deviations, with the mixing angle not running at all. Atomic parity-violation experiments also have sensitivity to physics beyond the SM, and in particular to speculative models with so-called leptoquarks or *additional* Z^0 bosons [71].

4.3 Parity Violation in One Trapped Ion

The value for the cesium weak charge measured by the Boulder group agrees to about one standard deviation with the prediction of the SM. However, this agreement came only after the facts, once the theory had been pushed to the level of accuracy called for by the experimental result. An independent high-precision measurement of the weak charge of cesium or of a different system, using a completely different experimental technique, would therefore be of great value, and several groups are pursuing this goal, cf. [65]. A novel experimental approach for cesium ($Z = 55$) atoms has been pioneered by the ENS-Paris group; plans exist at SUNY, Stony Brook, and at INFN, Legnaro, to measure the weak charge of francium ($Z = 87$), the heaviest alkali, in an atom trap; an ytterbium ($Z = 70$) experiment is planned at Berkeley.

A new way to measure parity violation was proposed by Fortson [72]. It exploits the ultrahigh sensitivity possible in experiments with *one single trapped ion*. Advantages that such experiments with one ion offer are the long coherence times of the atomic states and a good control of systematic errors. These can make up for higher counting rates in the more traditional atomic beam experiments. Several crucial steps toward a parity experiment have been demonstrated in pioneering experiments in Seattle with one $^{138}\text{Ba}^+$ ion [73, 74, 75]. Parity violation due to Z^0

exchange in an atom or ion results in a mixing of opposite-parity levels, primarily the S and P states. A small admixture of the P state into the S state makes a dipole (E1) transition to the first excited D state possible, where previously only an electric quadrupole (E2) transition was allowed by parity selection rules. From the interference between the two amplitudes the parity-violating amplitude can be extracted, and from it the weak charge Q_w of the ion. In the case of Ba$^+$, the $6S_{1/2}$–$5D_{3/2}$ E2 transition, which is parity-allowed, acquires a small parity-violating induced E1 amplitude, due to the small admixture of the $6P_{1/2}$ state into the $6S_{1/2}$ state. The interference of this parity-forbidden E1 amplitude with the parity-allowed E2 amplitude is observable by measuring the light-shift of the two ground-state Zeeman levels.

The radium ion, Ra$^+$, is the heaviest of the earth-alkalis ($Z = 88$) and has similar electronic structure as Ba$^+$ (and the cesium and francium atoms), see Fig. 4. Since parity-violation effects in atomic systems grow faster than Z^3 [63], radium will have much larger parity violation than barium: The relevant parity-violating $S_{1/2}$–$D_{3/2}$ amplitude is, in fact, expected to be about a factor of 20 larger in the Ra$^+$ ion than in the cesium atom or the Ba$^+$ ion [76], resulting in relatively smaller systematic errors and a larger statistical sensitivity. Also, the relevant S–D transitions for Ra$^+$ are better suited for present-day lasers, compared to Ba$^+$. Therefore, the radium ion appears, on paper, the superior candidate for a high-precision atomic parity-violation experiment. Such an experiment is being set up at KVI. An important issue that needs to be studied is the optimal isotope for a parity-violation experiment. Odd isotopes would offer the advantage of using the hyperfine structure for eliminating the E2 transition in favor of a very weak magnetic dipole (M1) transition. In that case, the interference between this M1 amplitude and the parity-violating E1 amplitude would be measured. Finally, the S–D transition in the Ra$^+$ ion is also an excellent candidate for an ultraprecise atomic clock [77], which could offer an alternative for the famous state-of-the-art Hg$^+$ clock. Such precise clocks are of particular interest for scientific measurements and for society, e.g., for navigation at precise scales.

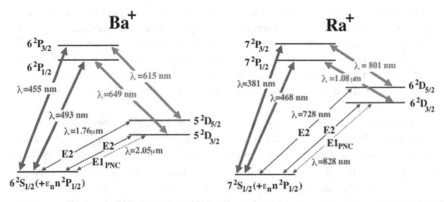

Fig. 4 The levels of the ^{138}Ba$^+$ (*left*) and ^{226}Ra$^+$ (*right*) ions in comparison. The wavelengths are indicated

Acknowledgments It is a pleasure to thank Frank Herfurth, Klaus Blaum, and Jürgen Kluge for organizing such an interesting and lively school at such a nice place and the Ph.D. students and postdocs for showing so much interest. I am also grateful to my experimental colleagues at KVI, in particular Klaus Jungmann, Gerco Onderwater, Lorenz Willmann, and Hans Wilschut, for many discussions on the topics addressed in this chapter. Finally, I would like to thank Cheng-Pang Liu for a pleasant collaboration.

References

1. M. Veltman, Reflections on the Higgs System, CERN 97-05, Geneva (1997).
2. R.N. Cahn, Rev. Mod. Phys. **68**, 951 (1996).
3. M.N. Harakeh et al., NuPECC long-range plan 2004, available from http://www.nupecc.org.
4. T. Akesson et al., Eur. Phys. J. C **51**, 421 (2007); hep-ph/0609216.
5. A.D. Sakharov, JETP Lett. **5**, 24 (1967).
6. C. Jarlskog, Phys. Rev. Lett. **55**, 1039 (1985).
7. P. Herczeg, Prog. Part. Nucl. Phys. **46**, 413 (2001).
8. G.D. Sprouse and L.A. Orozco, Annu. Rev. Nucl. Part. Sci. **47**, 429 (1997).
9. N.D. Scielzo et al., Phys. Rev. Lett. **93**, 102501 (2004).
10. A. Gorelov et al., Phys. Rev. Lett. **94**, 142501 (2005).
11. N. Severijns, M. Beck, and O. Naviliat-Cuncic, Rev. Mod. Phys. **78**, 991 (2006).
12. H.W. Wilschut and K.P. Jungmann, Nucl. Phys. News **17**, 11 (2007).
13. I.B. Khriplovich and S.K. Lamoreaux, CP Violation Without Strangeness: Electric Dipole Moments of Particles, Atom, and Molecules, Springer, Berlin (1997).
14. W. Fischler, S. Paban, and S. Thomas, Phys. Lett. B **289**, 373 (1992).
15. M. Pospelov and A. Ritz, Ann. Phys. (NY) **381**, 119 (2005).
16. C.-P. Liu, M.J. Ramsey-Musolf, W.C. Haxton, R.G.E. Timmermans, and A.E.L. Dieperink, Phys. Rev. C **76**, 035503 (2007).
17. L.I. Schiff, Phys. Rev. **132**, 2194 (1963).
18. P.G.H. Sandars, Contemp. Phys. **42**, 97 (2001).
19. J.S.M. Ginges and V.V. Flambaum, Phys. Rep. **397**, 63 (2004).
20. B.C. Regan, E.D. Commins, Ch.J. Schmidt, and D. DeMille, Phys. Rev. Lett. **88**, 071805 (2002).
21. A.N. Petrov et al., Phys. Rev. A **72**, 022505 (2005).
22. J.R. Guest et al., Phys. Rev. Lett. **98**, 093001 (2007).
23. M.V. Romalis, W.C. Griffith, J.P. Jacobs, and E.N. Fortson, Phys. Rev. Lett. **86**, 2505 (2001).
24. C.A. Baker et al., Phys. Rev. Lett. **97**, 131801 (2006).
25. F.J.M. Farley et al., Phys. Rev. Lett. **93**, 052001 (2004).
26. Yu.F. Orlov, W.M. Morse, Y.K. Semertzidis, Phys. Rev. Lett. **96**, 214802 (2006).
27. C.-P. Liu and R.G.E. Timmermans, Phys. Rev. C **70**, 055501 (2004).
28. C.-P. Liu and R.G.E. Timmermans, Phys. Lett. B **634**, 488 (2006).
29. R.N. Mohapatra et al., Theory of Neutrinos: A White Paper, arXiv:hep-ph/0510213.
30. Y. Fukuda et al. (SuperKamiokande), Phys. Rev. Lett. **81**, 1562 (1998).
31. Q.R. Ahmad et al. (SNO), Phys. Rev. Lett. **89**, 011301 (2002).
32. S.N. Ahmed et al. (SNO), Phys. Rev. Lett. **92**, 181301 (2004).
33. K. Eguchi et al. (KamLAND), Phys. Rev. Lett. **90**, 021802 (2003).
34. T. Araki et al. (KamLAND), Phys. Rev. Lett. **94**, 081801 (2005).
35. A.A. Aguilar-Arevalo et al. (MiniBooNE), Phys. Rev. Lett. **98**, 231801 (2007).
36. J.N. Bahcall and R. Davis, Jr., Science **191**, 264 (1976).
37. R. Davis, Jr., Rev. Mod. Phys. **75**, 985 (2003).
38. J.N. Bahcall, M.C. Gonzalez-Garcia, and C. Pena-Garay, Phys. Rev. Lett. **90**, 131301 (2003).
39. M. Goldhaber, L. Grodzins, and A.W. Sunyar, Phys. Rev. **109**, 1015 (1958).
40. A. de Gouvea, B. Kayser, and R.N. Mohapatra, Phys. Rev. D **67**, 053004 (2003).

41. E. Majorana, Nuovo Cim. **14**, 171 (1937).
42. W.C. Haxton and G.J. Stephenson, Jr., Prog. Part. Nucl. Phys. **12**, 409 (1984).
43. Y. Farzan and A.Yu. Smirnov, Phys. Lett. B **557**, 224 (2003).
44. Ch. Kraus et al., Eur. Phys. J. C **40**, 447 (2005).
45. S. Hannestad, Annu. Rev. Nucl. Part. Sci. **56**, 137 (2006).
46. D.N. Schramm, Comments Nucl. Part. Phys. **17**, 239 (1987).
47. T.J. Loredo and D.Q. Lamb, Phys. Rev. D **65**, 063002 (2002).
48. A. Osipowicz et al., KATRIN Letter of Intent, arXiv:hep-ex/0109033.
49. S. Elliott and P. Vogel, Annu. Rev. Nucl. Part. Phys. **52**, 115 (2002).
50. H.V. Klapdor-Kleingrothaus et al., Eur. Phys. J. A **12**, 147 (2001).
51. K. Zuber, arXiv:nucl-ex/0610007.
52. H.V. Klapdor-Kleingrothaus et al., Mod. Phys. Lett. A **16**, 2409 (2001).
53. J. Schechter and J.W.F. Valle, Phys. Rev. D **25**, 2951 (1982).
54. J.N. Bahcall, H. Murayama, and C. Pena-Garay, Phys. Rev. D **70**, 033012 (2004).
55. S. Rakers et al., Phys. Rev. C **70**, 054302 (2004).
56. S. Rakers et al., Phys. Rev. C **71**, 054313 (2005).
57. M. Fukugita and T. Yanagida, Phys. Lett. **174B**, 45 (1986).
58. W. Buchmuller, R.D. Peccei, and T. Yanagida, Annu. Rev. Nucl. Part. Sci. **55**, 311 (2005).
59. S. Weinberg, Phys. Rev. Lett. **43**, 1571 (1979).
60. A. Broncano, M.B. Gavela, and E. Jenkins, Phys. Lett. B **522**, 177 (2003).
61. A. Broncano, M.B. Gavela, and E. Jenkins, Nucl. Phys. **B672**, 163 (2003).
62. A. Czarnecki and W.J. Marciano, Nature **435**, 437 (2005).
63. M.-A. Bouchiat and C. Bouchiat, Phys. Lett. **48B**, 111 (1974); J. Phys. (Paris) **35**, 899 (1974); J. Phys. (Paris) **36**, 483 (1974).
64. I.B. Khriplovich, Parity Nonconservation in Atomic Phenomena, Gordon & Breach, New York (1991).
65. J. Guéna, M. Lintz, and M.-A. Bouchiat, Mod. Phys. Lett. **A20**, 375 (2005).
66. P.L. Anthony et al. (SLAC E158 Collaboration), Phys. Rev. Lett. **92**, 181602 (2004).
67. P.L. Anthony et al. (SLAC E158 Collaboration), Phys. Rev. Lett. **95**, 018601 (2005).
68. G.P. Zeller et al. (NuTeV Collaboration), Phys. Rev. Lett. **88**, 091802 (2002).
69. C.S. Wood, S.C. Bennett, D. Cho, B.P. Masterson, J.L. Roberts, C.E. Tanner, and C.E. Weiman, Science **275**, 1759 (1997).
70. S.C. Bennett and C.E. Weiman, Phys. Rev. Lett. **82**, 2484 (1999).
71. W.J. Marciano and J.L. Rosner, Phys. Rev. Lett. **65**, 2963 (1990).
72. N.E. Fortson, Phys. Rev. Lett. **70**, 2383 (1993).
73. T.W. Koerber, M.H. Schacht, K.R.G. Hendrickson, W. Nagourney, and E.N. Fortson, Phys. Rev. Lett. **88**, 143002 (2002).
74. T.W. Koerber, M. Schacht, W. Nagourney, and E.N. Fortson, J. Phys. B **36**, 637 (2003).
75. J.A. Sherman, T.W. Koerber, A. Markhotok, W. Nagourney, and E.N. Fortson, Phys. Rev. Lett. **94**, 243001 (2005).
76. V.A. Dzuba, V.V. Flambaum, and J.S.M. Ginges, Phys. Rev. A **63**, 062101 (2001).
77. B.K. Sahoo, B.P. Das, R.K. Chaudhuri, D. Mukherjee, R.G.E. Timmermans, and K. Jungmann, Phys. Rev. A **76**, 040504 (2007).

Principles of Ion Traps

G. Werth

1 Trapping Principles

Three-dimensional confinement of charged particles requires a potential energy minimum at some point in space in order that the corresponding force is directed toward that point in all three dimensions. In general, the dependence of the magnitude of this force on the coordinates can have an arbitrary form; however, it is convenient to have a binding force that is harmonic, since this simplifies the analytical description of the particle motion. Thus, we assume

$$F \propto -r \,. \tag{1}$$

It follows from

$$F = -\mathrm{grad}\, U \,, \tag{2}$$

where $U = Q\Phi$ is the potential energy, that in general the required function Φ is a quadratic form in the Cartesian coordinates x, y, z:

$$\Phi = \frac{\Phi_0}{d^2}(Ax^2 + By^2 + Cz^2) \,, \tag{3}$$

where A, B, and C are some constants, d a normalizing factor, and Φ_0 can be a time-dependent function. If we attempt to achieve such confinement using an electrostatic field acting on an ion of charge Q, we find that in order to satisfy Laplace's equation $\Delta\Phi = 0$, the coefficients must satisfy $A + B + C = 0$. For the interesting case of rotational symmetry around the z-axis, this leads to $A = B = 1$ and $C = -2$, giving us the quadrupolar form

$$\Phi = \frac{\Phi_0}{d^2}(x^2 + y^2 - 2z^2) = \frac{\Phi_0}{d^2}(\rho^2 - 2z^2) \,, \tag{4}$$

G. Werth
Johannes Gutenberg University, D-55099 Mainz, Germany

Werth, G.: *Principles of Ion Traps.* Lect. Notes Phys. **749**, 31–67 (2008)
DOI 10.1007/978-3-540-77817-2_2

with $\rho^2 = x^2 + y^2$. If the radial distance from the center ($\rho = z = 0$) of a hyperbolic trap to the ring electrode is called r_0, and the axial distance to an endcap is z_0, the equations for the hyperbolic electrode surfaces are

$$\rho^2 - 2z^2 = r_0{}^2 , \tag{5}$$
$$\rho^2 - 2z^2 = -2z_0^2.$$

If the potential difference between the ring and endcaps is taken to be ϕ_0, then

$$d^2 = r_0^2 + 2z_0{}^2 . \tag{6}$$

From the difference in signs between the radial and axial terms, we see that the potential has a saddle point at the origin, having a minimum along one coordinate but a maximum along the other. Earnshaw's theorem states that it is not possible to generate a minimum of the electrostatic potential in free space. Nevertheless, it is possible to circumvent Earnshaw's theorem by superimposing a magnetic field along the z-axis to create what is called the *Penning trap* or to use time-dependent electric field, leading to the *Paul trap* [1, 2].

The electrodes which create a quadrupole potential consist of three hyperbolic sheets of revolution: a ring electrode and two endcaps (Fig. 1) which share the same asymptotic cone. The size of the device ranges, in different applications, from several centimeters for the characteristic dimension d to fractions of a millimeter. The trapped charged particles are constrained to a very small region of the trap, whose position can be centered by using a small additional dc field.

In recent years, different trap geometries have become common which are easier to manufacture and align and in addition allow optical access to the trapped particles without further modification (Fig. 2): the *linear Paul trap* which uses four parallel rods as electrodes. An ac voltage applied between adjacent electrodes leads to dynamical confinement similar as in the three-dimensional case. Axial confinement is provided by a static voltage at end electrodes. Figure 2a shows an example. For static trapping, *cylindrical Penning traps* are in use (Fig. 2b).

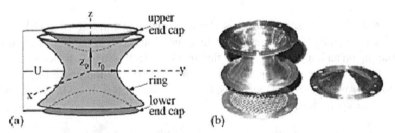

Fig. 1 Basic arrangement for Paul and Penning traps (**a**). The inner electrode surfaces are hyperboloids. The dynamic stabilization in the Paul trap is given by an ac voltage $V_0 \cos \Omega t$. The static stabilization in the Penning trap is given by a dc voltage $U = U_0$ and an axial magnetic field. (**b**) A photograph of a trap with $\rho = 1$ cm is given. One of the endcaps is formed as mesh to allow optical access to trapped ions

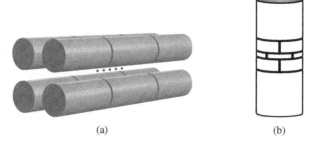

Fig. 2 Linear Paul trap (**a**) and open endcap cylindrical Penning trap (**b**). A radio-frequency field applied to the rods of the linear Paul trap confines charged particles in the radial direction, a dc voltage at the end segments serves for radial trapping. The Penning trap has guard electrodes between the central ring and the endcaps to compensate partly for deviations from the ideal quadrupole potential near the trap center. Ion excitation can be performed by rf fields applied to segments of the electrodes

These electrode geometries produce a harmonic binding force near the center precisely of the classical form. Further away from the origin, higher order components in the potential will become significant. For cylindrical Penning traps, these can be partially reduced by additional compensation electrodes placed between the ring and endcaps as shown in Fig. 2.

1.1 Paul Traps

General Principles

In an ideal Paul trap, an oscillating electric potential usually in combination with a static component, $U_0 + V_0 \cos \Omega t$, is applied between the ring and the pair of endcap electrodes. It creates a potential of the form

$$\Phi = \frac{U_0 + V_0 \cos \Omega t}{2d^2}(r^2 - 2z^2) \,. \tag{7}$$

Since the trapping field is inhomogeneous, the average force acting on the particle, taken over many oscillations of the field, is not zero. Depending on the amplitude and frequency of the field, the net force may be convergent toward the center of the trap leading to confinement or divergent leading to the loss of the particle. Thus, although the electric force alternately causes convergent and divergent motion of the particle in any given direction, it is possible by appropriate choice of field amplitude and frequency to have a time-averaged restoring force in all three dimensions toward the center of the trap as required for confinement [3].

The conditions for stable confinement of an ion with mass M and charge Q in the Paul field may be derived by solving the equation of motion:

$$\frac{\partial^2 u}{\partial t^2} = \frac{Q}{Md^2}(U_0 + V_0 \cos \Omega t)u , \tag{8}$$

$$u = x, y, z.$$

Using the dimensionless parameters $a_x = a_y = -2a_z = -4QU_0/Md^2\Omega^2$ and $q_x = q_y = 2q_z = 2QV_0/Md^2\Omega^2$, we obtain a system of three differential equations of the homogeneous Mathieu type [4, 5]:

$$\frac{\partial^2 u}{\partial t^2} + (a - 2q\tau)u = 0 , \tag{9}$$

where $\tau = \Omega t$.

The values of a and q for which the solutions are stable simultaneously for both directions, an obvious requirement for three-dimensional confinement, are found by using the relationship $a_z = -2a_r$ and $q_z = 2q_r$ to make a composite plot of the boundaries of stability for both directions on the same set of axes (Fig. 3): the overlap regions lead to three-dimensional confinement. The most important for practical purposes is the stable region near the origin which has been exclusively used for ion confinement.

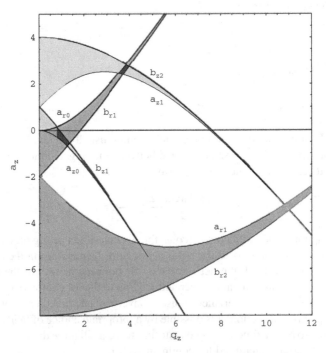

Fig. 3 The stability domains for the ideal Paul trap. *Light gray*: z-direction; *dark gray*: r-direction. Three-dimensional stability is assured in the overlapping regions

The stable solutions of the Mathieu equation can be expressed in the form of a Fourier series; thus

$$u_j(\tau) = A_j \sum_{n=-\infty}^{n=+\infty} c_{2n} \cos(\beta_j + 2n)\tau + B_j \sum_{n=-\infty}^{n=+\infty} c_{2n} \sin(\beta_j + 2n)\tau , \qquad (10)$$

where A_j and B_j are constants depending of the initial conditions. The stability parameter β are functions of a and q. The coefficients c_{2n} which are the amplitudes of the Fourier components of the particle motion, decrease with increasing n.

For small values of $a, q << 1$, we can approximate the stability parameter β by

$$\beta_i^2 \approx a_i + q_i^2/2 . \qquad (11)$$

In this so-called adiabatic approximation, the coefficients c_{2n} become rapidly smaller with increasing n. For $n = 1$, we have $c_{-2} = c_{+2} = -(q_i/4)c_0$. The ion motion simplifies to

$$u_i(t) = A \left(1 - \frac{q_i}{2} \cos \Omega t\right) \cos \omega_i t , \qquad (12)$$

with

$$\omega_i = \beta_i \Omega/2 . \qquad (13)$$

This can be considered as the motion of an oscillator of frequency ω whose amplitude is modulated with the trap's driving frequency Ω. Since it is assumed that $\beta << 1$, the oscillation at ω, usually called the secular or macromotion, is slow compared with the superimposed fast micromotion at Ω. Because of the large difference in the frequencies ω and Ω, the ion motion can be well separated into two components and the behavior of the slow motion at frequency ω can be considered as separate, while time averaging over the fast oscillation at Ω. Wuerker et al. [6] have taken photographs of single particle trajectories of microparticles in a Paul-type trap at low frequencies of the trapping field demonstrating the validity of this approximation (Fig. 4).

Fig. 4 Observed trajectory of a microparticle in a Paul-type trap [6]

Potential Depth

In the adiabatic approximation, an expression for the depth of the confining potential can be derived when we consider the secular motion only. The ion behaves in the axial direction as an harmonic oscillator of frequency ω_z. For no dc voltage ($a = 0$), we have

$$\omega_z = \frac{QV_0}{\sqrt{2}Mz_0^2\Omega}. \tag{14}$$

This corresponds to a time-averaged (pseudo)potential of depth in the axial direction of

$$\bar{D}_z = \frac{QV_0^2}{4Mz_0^2\Omega^2}, \tag{15}$$

and similarly in the radial direction,

$$\bar{D}_r = \frac{QV_0^2}{4Mr_0^2\Omega^2}. \tag{16}$$

For $r_0^2 = 2z_0^2$, we have $\bar{D}_r = \bar{D}_z/2$.

The effect of an additional dc voltage on the trap electrodes is to alter the depth of the potential in the field direction. If the voltage U_0 is applied symmetrically to the trap electrodes (that is, the trap center is at zero potential), we have

$$\begin{aligned}
\bar{D}'_z &= \bar{D}_z + U_0/2, \\
\bar{D}'_r &= \bar{D}_r - U_0/2.
\end{aligned} \tag{17}$$

Typical values for a trap of 1 cm radius r_0 driven at a frequency of $\Omega/2\pi = 1\,\text{MHz}$ and an ion atomic mass of 100 are 15 eV for the radial potential depth.

Motional Spectrum

Equation (10) shows that the motional spectrum contains the frequencies $(\beta_j + 2n)\Omega$, n integer, with the fundamental frequencies given by $n = 0$. To experimentally demonstrate this spectrum, the motion can be excited by an additional (weak) rf field applied to the electrodes. When resonance occurs between the detection field frequency and one of the frequencies in the ion spectrum, the motion becomes excited and some ions may leave the trap, providing a signal commensurate with the number of trapped ions. Figure 5 shows an example where a cloud of stored N_2^+ ions is excited at resonances occurring at the frequencies predicted by theory.

Optimum Trapping Conditions

When we define the optimum trapping conditions for a Paul trap as those which yield the highest density of trapped particles, we are faced with conflicting

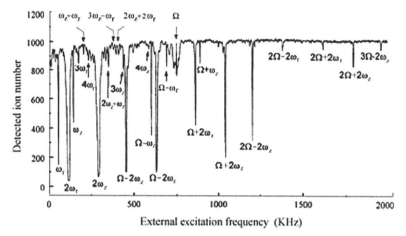

Fig. 5 The spectrum of motional resonances of an N_2^+ ion cloud in a quadrupole Paul trap (from [7])

requirements: On the one hand, the maximum trapped ion number increases with the potential depth D, since the number is limited by the condition that the space charge potential of the ions not exceed D. On the other hand, the oscillation amplitudes also increase at the same time, resulting in increasing ion loss for higher q in a trap of a given size. Consequently, the maximum ion number is expected in a region around the center of the stability diagram. This has been confirmed experimentally by systematic variation of the trapping parameters and a measurement of the relative trapped ion number by laser-induced fluorescence [8] (Fig. 6).

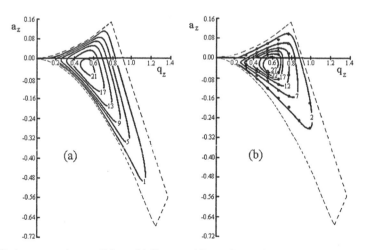

Fig. 6 Optimum trapping conditions. (**a**) Computed lines of equal ion density within the stability diagram. The numbers give relative densities; (**b**) experimental lines of equal ion density from laser-induced fluorescence (from [8])

Storage Time

Once the ions are stored in a Paul trap, they would remain there in the ideal case for infinite long time. In practice, the storage time may be limited by trap imperfections as discussed below. They can be avoided by proper choice of the operating conditions. The limiting factor then would be collisions with neutral background molecules. Thus, operation of the traps in ultra-high vacuum seems necessary. In fact storage times for atomic ions of many hours are obtained routinely when working at pressures around 10^{-10} mbar.

Under certain conditions, however, higher background pressure may have a beneficial effect on the storage time as well. Major and Dehmelt [9] have pointed out that ion-neutral collisions lead to damping of the ion motion and thus to increased storage times when the ion mass exceeds the mass of the neutral atom or molecule. This has been experimentally verified in many cases, and storage times of many days for heavy ions such as Ba^+ or Pb^+ have been obtained when operating with light buffer gases (He, N_2, Ne) at pressures as high as 10^{-4} mbar.

Ion Density Distribution

In thermal equilibrium at high temperatures, a cloud of trapped ions assumes a Gaussian density distribution, averaged over a period of the micromotion. This has been experimentally confirmed by scanning a laser spatially through the trap and observing the fluorescence light after laser excitation (Fig. 7). When the ion temperature is lowered, the cloud diameter shrinks until a homogeneously charged sphere is obtained.

From the width of the distribution, the ions' average kinetic energy can be derived [10]. As a rule of thumb, it amounts to 1/10 of the potential depth.

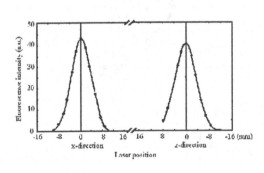
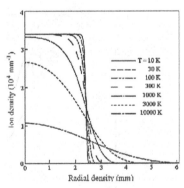

Fig. 7 *Left*: Measured density distribution of Ba^+ ions in axial and radial directions in a Paul trap of 4 cm ring radius showing a Gaussian density distribution (from [10]). *Right*: Calculated density distributions for different ion temperatures (from [11])

Storage Capability

The maximum density n of ions that can be stored is given when the space charge potential $V_{sc} = 4\pi\rho = 4\pi en$ created by the ion cloud equals the trap potential depth. For a spherical potential shape, it follows that

$$n = 3\bar{D}/4\pi e^2 r_0^2 \,. \tag{18}$$

The total stored ion number N follows from integration over the active trap volume. Taking a density distribution as shown in Fig. 7, we arrive at a maximum ion number $N \approx 10^6$ in fair agreement with experimental observations.

Influence of Paul Trap Imperfections

A single ion in a perfect quadrupole potential does not describe a real experimental situation. Truncations of the electrodes, misalignments, or machining errors change the shape of the potential field. The equations of motions as discussed above are valid only for a single confined particle. Simultaneous confinement of several particles requires the consideration of space charge effects. For a low-density cloud, it seems reasonable to assume that the particles move in somewhat modified orbits, mainly independent of each other except for rare Coulomb scattering events. These collisions ultimately serve to establish a thermal equilibrium between the particles. They may be considered as small perturbations to the particle motion, provided the time average of the Coulomb interaction potential is small compared to the average energy of the individual particle. Then the particle cloud may be described as ideal gas of non-interacting particles in thermal equilibrium.

Deviations of the trap potential from the ideal quadrupolar form can be treated by a series expansion in spherical harmonics; thus

$$\Phi(\rho, \theta) = (U_0 + V_0 \cos \Omega t) \sum_{n=2}^{\infty} c_n \left(\frac{\rho}{d}\right)^n P_n(\cos \theta) \,, \tag{19}$$

where $P_n(\cos \theta)$ are the Legendre polynomials of order n. For rotational symmetry, the odd coefficients c_n vanish. The terms beyond the quadrupole (c_2) may be looked on as perturbing potentials, the lowest of which is the octupole (c_4), followed by the dodecapole (c_6). The equations of motion for a single particle in an imperfect Paul trap now become coupled inhomogeneous differential equations which cannot be solved analytically. It has been shown, however, that, under certain conditions that would otherwise give stability in a perfect quadrupole field, the motion becomes unstable [12, 13]. These conditions can be expressed in terms of the stability parameters β_r and β_z; thus

$$n_r \beta_r + n_z \beta_z = 2k \,, \tag{20}$$

or equivalently

$$n_r \omega_r + n_z \omega_z = k\Omega \,, \tag{21}$$

where n_r, n_z, and k are integers. This relationship states that if a linear combination of harmonics of the ion macrofrequencies coincides with a harmonic of the high-frequency trapping field, an ion will gain energy from that field until it gets lost from the trap.

Experimental proof of the instabilities has been obtained by measurements of the number of trapped ions at different operating points. A high-resolution scan of the stability diagram was given by Alheit et al. [14], who observed instabilities resulting from very high orders of perturbing potentials (Fig. 8). From this figure, it is evident that strong instabilities occur at the high-q region of the stability diagram, due to hexapole and octupole terms in the expansion of the potential, which are the highest orders expected in a reasonably well-machined trap. This makes it very difficult to

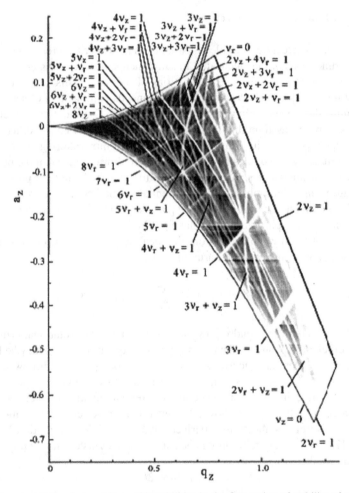

Fig. 8 Experimentally observed lines of instabilities in the first region of stability of a real Paul trap taken with H_2^+ ions. The unstable lines are assigned according to (20). The intensity of gray is proportional to the trapped ion number (from [14])

obtain long storage times at high amplitudes of the trapping voltage (for a given frequency). In fact, the most stable conditions are obtained for small q values near the $a = 0$ axis; here the instability condition on the frequencies is met only for a very high resonance order which would occur only if the field was highly imperfect.

As another consequence of the presence of the higher order terms in the trapping potential, the motional eigenfrequencies are shifted with respect to the pure quadrupole field, and moreover, in an amplitude-dependent way. These shifts are of particular importance in the case of Penning traps, when they are used for very high resolution mass spectrometry, and will be discussed later. While for the ideal harmonic potential the line shape of the resonances is a Lorentzian, it now becomes asymmetric. From a fit to the line, the size of the higher order coefficients in the series expansion of the potential (19) can be determined [15]. In addition to the ideal case, the motional spectrum now contains not only the eigenfrequencies as calculated from (10) but also the sum and difference frequencies.

1.2 Penning Traps

Theory of the Ideal Penning Trap

The Penning trap uses static electric and magnetic fields to confine charged particles. The ideal Penning trap is formed by the superposition of a homogeneous magnetic field $B = (0,0,B_0)$ and an electric field $E = -\nabla\Phi$ derived from the quadrupole potential as given in (4).

A particle of mass M, charge Q, and velocity $v = (v_x, v_y, v_z)$ moving in the fields E and B experiences a force

$$\vec{F} = -Q\Phi + Q(\vec{v} \times \vec{B}) . \tag{22}$$

Since the magnetic field is along the z-axis, the z-component of the force is purely electrostatic, and therefore to confine the particle in the z-direction we must have $QU > 0$. The x- and y-components of F are a combination of a dominant restraining force due to the magnetic field, characterized by the cyclotron frequency

$$\omega_c = \frac{|QB_0|}{M} \tag{23}$$

and a repulsive electrostatic force that tries to push the particle out of the trap in the radial direction. Newton's equations of motion in Cartesian coordinates are as follows:

$$\frac{d^2x}{dt^2} - \omega_c \frac{dy}{dt} - \frac{1}{2}\omega_z^2 x = 0,$$
$$\frac{d^2y}{dt^2} + \omega_c \frac{dx}{dt} - \frac{1}{2}\omega_z^2 y = 0 , \tag{24}$$

$$\frac{d^2z}{dt^2} + \omega_z{}^2 z = 0,$$

$$\omega_z = \sqrt{\frac{2QU_0}{Md^2}}.$$ (25)

The motion in the z-direction is a simple harmonic oscillation with an axial frequency ω_z decoupled from the transverse motion in the x and y directions.

To describe the motion in the x, y plane, we introduce a complex variable $u = x + iy$. The radial equations of motion then reduce to

$$\frac{d^2u}{dt^2} + i\omega_c \frac{du}{dt} - 2\omega_z{}^2 u = 0.$$ (26)

The general solution is given by

$$u(t) = R_+ e^{-i(\omega_+ t + \alpha_+)} + R_- e^{i(\omega_- t + \alpha_-)},$$ (27)

with

$$\omega_+ = \frac{1}{2}\left(\omega_c + \sqrt{\omega_c{}^2 - 2\omega_z{}^2}\right),$$ (28)

$$\omega_- = \frac{1}{2}\left(\omega_c - \sqrt{\omega_c{}^2 - 2\omega_z{}^2}\right).$$ (29)

ω_+ is called the modified cyclotron frequency and ω_- the magnetron frequency. R_+, R_-, α_+, α_- are the radii and the phases of the respective motions, determined by the initial conditions. Equation (27) is the equation of an epicycloid. It describes a trajectory around the trap center where the particle moves within a circular strip between an outer radius $R_+ + R_-$ and an inner radius $|R_+ - R_-|$. An example is shown in Fig. 9. Figure 10 shows a sketch of the ion trajectory in all three dimensions.

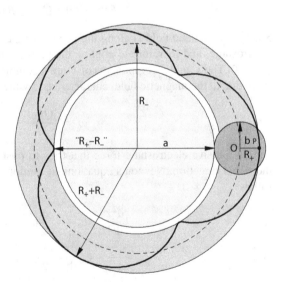

Fig. 9 Radial motion of a trapped particle according to (27) for the parameters $\omega_+ = 4\omega_-$, $R_+ = 3R_-$, $\alpha_+ = \alpha_- = 0$

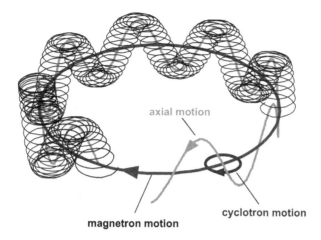

axial motion

cyclotron motion

magnetron motion

Fig. 10 Sketch of the ion trajectory in a Penning trap

In order that the motion can be bounded, the roots of (28) and (29) must be real, leading to the trapping condition

$$\omega_0{}^2 - 2\omega_z{}^2 > 0 \,, \tag{30}$$

or equivalently

$$\frac{|Q|}{M}B_0^2 > \frac{4\,|U_0|}{d^2}, \quad QU_0 > 0\,. \tag{31}$$

Several useful relations exist between the eigenfrequencies of the trapped particle:

$$\omega_+ + \omega_- = \omega_c \,, \tag{32}$$

$$2\omega_+\omega_- = \omega_z{}^2 \,, \tag{33}$$

$$\omega_+{}^2 + \omega_-{}^2 + \omega_z{}^2 = \omega_c{}^2 \,. \tag{34}$$

Motional Spectrum in Penning Trap

The motional spectrum of an ion in a Penning trap contains the fundamental frequencies ω_+, ω_-, and ω_z as shown in the solution of the equations of motion. These frequencies can be measured using resonant excitation of the ion motion by applying an additional dipole rf field to the trap electrodes. An indication of resonant excitation is a rapid loss of stored ions due to the gain in energy from the rf field, which enables them to escape. The electrostatic field inside the trap is expected to be a superposition of many multipole components; hence many combinations of the fundamental frequencies and their harmonics become visible, depending on the amplitude of the exciting field. Figure 11 shows the number of trapped electrons as a function of the frequency of an excitation rf field which enter the apparatus by an antenna placed near the trap. Identification of the resonances is through their

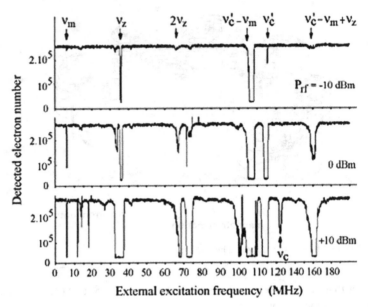

Fig. 11 Observed motional resonances of electrons in a Penning trap taken at different amplitudes of an exciting rf field

different dependence on the electrostatic field strength. The axial frequency ω_z depends on the square root of the trapping voltage while ω_+ and ω_- vary linearly with the voltage.

Of particular interest is the combination $\omega_+ + \omega_-$, since it is independent of the trapping voltage, and equals the cyclotron frequency of the free ion ω_c whose measurement can serve to calibrate the magnetic field at the ion position.

The frequency ω_c can be obtained in three different ways: direct excitation of the sideband $\omega_+ + \omega_-$, measurement of the fundamental frequency ω_+ at different trapping voltages U_0, and extrapolating to $U_0 = 0$, or using relation (34). The last is particularly useful since it is, to first order, independent of perturbations arising from field imperfections, which may shift the individual eigenfrequencies [16].

Penning Trap Imperfections

Similar as in the case of Paul traps, in real Penning traps, the conditions are considerably more complicated than the ideal description in the previous section. The hyperbolic surfaces are truncated and may deviate from the ideal shape, the trap axis may be tilted with respect to the direction of the magnetic field, and the long-range Coulomb potential of ions adds to the trapping potential when more than a single is trapped. The departures of the field from the pure quadrupole introduce non-linearity in the equations of motion, and coupling between the degrees of freedom.

The main effects of these imperfections on the behavior of the trapped ions are a shift of the eigenfrequencies and a reduction in the storage capability of the trap.

Shift of the Eigenfrequencies

(1) Electric Field Imperfections

As in the case of the previously discussed Paul trap, imperfections in the electrostatic field can be treated by a multipole expansion of the potential with coefficients c_n characterizing the strength of higher order components in the trapping potential as stated in (19). Several authors [17, 18, 19, 20] have calculated the resulting shift in the ion oscillation frequency, assuming that due to rotation symmetry of the trap only even orders contribute. Restriction to the octupole term (c_4) as most important contribution, the results are

$$\Delta\omega_z = \frac{3}{4}\frac{c_4}{d^2}\omega_z \left[R_z^{\,2} - 2\left(R_+^{\,2} + R_-^{\,2}\right)\right],$$

$$\Delta\omega_+ = \frac{3}{2}\frac{c_4}{d^2}\frac{\omega_z^{\,2}}{\omega_+ - \omega_-}\left[R_+^{\,2} + 2R_-^{\,2} - 2R_z^{\,2}\right], \tag{35}$$

$$\Delta\omega_- = -\frac{3}{2}\frac{c_4}{d^2}\frac{\omega_z^{\,2}}{\omega_+ - \omega_-}\left[R_-^{\,2} + 2R_+^{\,2} - 2R_z^{\,2}\right].$$

In order to minimize these shifts, it is obviously necessary not only to make the trap as perfect as possible but also to reduce the radii R by some ion cooling method.

(2) Magnetic Field Inhomogeneities

An inhomogeneity in the superimposed magnetic field B_0 of the Penning trap also leads to a shift of the field-dependent frequencies. As in the electric field case, a magnetostatic field in a current-free region can be derived from a potential function Φ_m, which can be expanded in the following multipole series:

$$\Phi_m = B_0 \sum_{n=0}^{\infty} b_n \rho^n P_n(\cos\theta). \tag{36}$$

In the case of mirror symmetry in the trap's midplane, the odd coefficients vanish. If we retain only the b_2 term in the expansion, the solution of the equations of motion leads to a frequency shift for the most interesting case of the sideband $\omega_+ + \omega_- = \omega_c$ given by

$$\Delta\omega_c = \omega_c\frac{b_2}{2}\left(R_z + \frac{\omega_-R_+ - \omega_+R_-}{\omega_+ - \omega_-}\right). \tag{37}$$

Other imperfections such as an ellipticity of the trap or a tilt angle between the trap axis and the magnetic field direction also shift the eigenfrequencies. It should be noted, however, that these shifts do not affect the free ions' cyclotron frequency ω_c in first order when using the "invariance theorem" $\omega_c^{\,2} = \omega_+^{\,2} + \omega_-^{\,2} + \omega_z^{\,2}$ [16].

(3) Space Charge Shift

The space charge potential of a trapped ion cloud also shifts the eigenfrequencies of individual ions. From a simple model of a homogeneous charge in the cloud, we are led to a decrease in the axial frequency proportional to the square root of the ion number. This has been verified experimentally (Fig. 12). More important for high-precision spectroscopy is the space charge shift in the frequencies when we are dealing with small numbers of ions, where the simple model is no longer adequate. The observed shift depends of the mean inter-ion distance and consequently on the ion temperature. For ions cooled to 4 K, Van Dyck et al. [21] have observed shifts of opposite signs in the perturbed cyclotron and magnetron frequencies, amounting to 0.5 ppb per stored ion. The shift increases approximately linearly with the charge of the ions. The electrostatic origin of the shifts suggests that the sum of the perturbed cyclotron and magnetron frequencies, which equals the free ion cyclotron frequency, should be independent of the ion number. In fact, the shifts observed in the sideband $\omega_+ + \omega_- = \omega_c$ are consistent with zero.

(4) Image Charges

An ion oscillating with amplitude r induces in the trap electrodes image charges which create an electric field that reacts on the stored ion and shifts its motional frequencies. The shift has been calculated by Van Dyck and coworkers [21] using a simple model. If the trap is replaced by a conducting spherical shell of radius a, the image charge creates an electric field given by

$$E = \frac{1}{4\pi\varepsilon_0} \frac{Qra}{(a^2 - r^2)^2} . \tag{38}$$

Since this field is added to the trap field, it causes a shift in the axial frequency amounting to

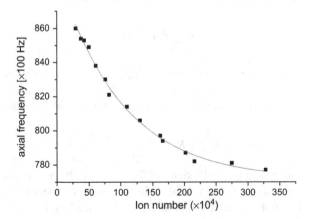

Fig. 12 Shift of the axial frequency as function of the trapped ion number

$$\Delta\omega_z = \frac{1}{4\pi\varepsilon_0} \frac{Q^2}{2Ma^3\omega_z} . \qquad (39)$$

It scales linearly with the ion number n and is significant only for small trap sizes. Since it is of purely electrostatic origin, the electric field-independent combination frequency $\omega_+ + \omega_- = \omega_c$ is not affected.

Instabilities of the Ion Motion

Another consequence of the presence of higher order multipoles in the field is that ion orbit instability can occur at certain operating points, where it would otherwise be stable in a perfect quadrupole field. Using perturbation theory to solve the equations of motions when the trap potential is written as a series expansion in spherical harmonics, Kretzschmar [17] has shown that the solution exhibits singularities for operating points at which

$$n_+\omega_+ + n_-\omega_- + n_z\omega_z = 0 , \qquad (40)$$

where n_+, n_-, n_z are integers. The ion trajectory at such points is unstable and the ion is lost from the trap. In Fig. 13, the results of measurements obtained for different trap voltages at a fixed value of the magnetic field on a cloud of trapped electrons are shown. The combined effect of space charge and trap imperfections leads to the loss of the particles [22]. The observed instabilities which become more discernible for extended storage times can be assigned to operating conditions predicted from (40).

These results show that in practice it is very difficult to obtain stable operating conditions when the trapping voltage exceeds about half the value allowed by the stability criterion.

Storage Time

Ideally, a stored ion would remain in a Penning trap for unlimited time. As discussed above, trap imperfections, however, may lead to ion loss. Proper choice of the operating conditions and reduced ion oscillation amplitude by ion cooling makes this negligible. The limiting factor then would be collisions with neutral background molecules since they tend to increase the magnetron orbit and eventually the ion will hit the trap boundary. Typical storage times at 10^{-10} mbar are several minutes, and at extremely low pressures of below 10^{-16} mbar, obtained by cryopumping, storage times of many months are obtained.

The problem of collision-induced increase of the magnetron orbit can be overcome by an additional rf field at the sum frequency of the perturbed cyclotron and the magnetron motion [23]. The field is applied between adjacent segments of the ring electrode of the Penning trap which is split into for quadrants. The effect of this field is to couple both motions. The cyclotron motion is damped by the collision with

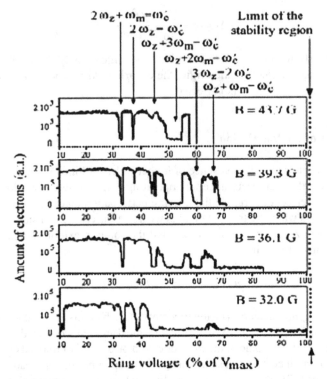

Fig. 13 Observed number of trapped electrons vs. trapping potential at different magnetic field strengths. The trap voltage is given in units of the maximum voltage as derived from (31). The combinations of eigenfrequencies leading to trap instabilities according to (40) are indicated (from [22])

background atoms, and the coupling transfers the damping to the magnetron motion. The net effect is that the ions aggregate near the trap center and consequently the storage time is extended [24].

Storage Capability

The actual behavior of the trapped particles, if there are more than one, will also be affected by their mutual electrostatic interaction. From the condition for stable confinement of individual particles, which gives in effect the magnetic field required to balance the radial component of the applied electric field, we can find a value for the limit on the number of ions that can be trapped at a given magnetic field intensity. This limit, sometimes called the Brillouin limit, is given by

$$n_{\lim} = B^2 / 2\mu_0 mc^2 , \tag{41}$$

where μ_0 is the permeability of free space. Naturally, as this limit is approached, the individual particle picture is no longer valid: one is then dealing with a plasma.

Spatial Distribution

Similar as in the Paul trap, an ion cloud in thermal equilibrium assumes a Gaussian density distribution. The width of the distribution, however, can be changed by a technique introduced in 1977 at the University of California at San Diego and called the "rotating wall" [25]: The crossed electric and magnetic field of a Penning trap causes a charged particle to rotate around the trap center at the magnetron frequency. In addition, the cloud may be forced to rotate at different frequencies by an external torque from an electric field rotating around the symmetry axis. The ring electrode of a Penning trap is split into different segments. A sinusoidal voltage V_{wj} of frequency ω_w is applied to the segments at $\theta_j = 2\pi j/n$ when n is the number of segments, with $V_{wj} = A_w \cos[m(\theta_j - \omega_w t)]$. Figure 14 illustrates an eight-segment configuration.

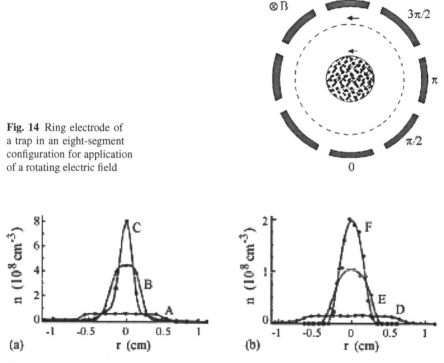

Fig. 14 Ring electrode of a trap in an eight-segment configuration for application of a rotating electric field

Fig. 15 Measured change of density by application of rotating wall field on an electron (**a**) and Be$^+$ (**b**) plasma. B,E: No rotating field, A,D: rotating field co-rotating with magnetron motion, C,F: rotating field counter-rotating with magnetron motion (from [26])

The additional centrifugal force from the plasma rotation leads to a change in plasma density and spatial profile. Thus, variation of the rotating wall frequency allows the plasma to compress or expand when the field is co-rotating or counter-rotating with the rotation of the plasma, respectively (Fig. 15).

2 Trap Techniques

2.1 Trap Loading

In-Trap Ion Creation

The easiest way to load ions into the trap is to create them inside the trapping volume by photo- or electro-ionization of an atomic beam or the neutral background gas. At room temperature, the energy of the ions is in general significantly smaller than the trap's potential depth, and all ions are confined. Alternatively, the ions can be created by surface ionization from a filament placed at the edge of the trapping volume (Fig. 16). Atoms of interest can be deposited on the filament surface or injected as ions into the material. Heating the filament releases ions from the surface. The efficiency depends on the work function of the filament material and the ionization potential of the atom under investigation. Using filaments of Pt, W, or Rh with high workfunction and atoms like Ba, Sr, and Ca with low ionization potential typical efficiencies are of the order of 10^{-5}.

Ion Injection from Outside

An ion injected into a Penning trap from an outside source under high vacuum condition may be captured by proper switching of the trap electrodes: When the ion

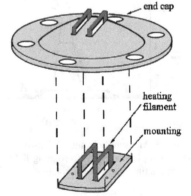

Fig. 16 Filaments for ion production by surface ionization placed in a slot of a trap's endcap electrode

travels along the magnetic field lines and approaches the first endcap electrode, its potential is set to zero, while the second endcap is held at some retarding potential. When the axial energy of the ion is smaller than the retarding potential it will be reflected. Before the ion leaves the trap through the first endcap, its potential is raised and prevents the ion to escape (Fig. 17).

This simple method requires that the arrival time of the ion at the trap is known, which can be achieved by pulsing the ion source. Moreover, the switching has to be performed in a time shorter than twice the transit time of the ions through the trap. For ion kinetic energies of a few 100 eV and a trap size of 1 cm, this time is of the order of 100 ns. Successful capture of ions from a pulsed source is routinely used at ISOLDE-facility at CERN, where nearly 100% efficiencies are achieved [27].

Injection into a Paul trap under similar conditions is more difficult, since the time-varying trapping potential of typically a MHz frequency cannot be switched from zero to full amplitude in a time of the order of less than a microsecond. The ion longitudinal kinetic energy, however, may be transferred into transverse components by the inhomogeneous electric trapping field and thus the ions may be confined for some finite time. Schuessler and Chun-sing [28] have made extensive simulations and phase considerations and have found that ions injected at low energy during a short interval when the ac trapping field has zero amplitude may remain in the trap for some finite time.

The situation is different when the ions undergo some kind of friction while they pass through the trap. This friction is most easily obtained by collisions with a light buffer gas. The density of the buffer gas has to be at least of such a value that the mean free pass of the ions between the collisions is of the trap's size. Coutandin and Werth [29] have shown that the trap is filled up to its maximum capacity in a short time at pressures around 10^{-3} Pa in a 1 cm size Paul trap when ions are injected along the trap axis with a few keV kinetic energy (Fig. 18). The same method is used routinely in many experiments where a high-energy ion beam is delivered from an accelerator. It is stopped and cooled by buffer gas collisions in a gas-filled linear radio-frequency trap and then ejected at low energies and with small energy spreads into a trap operated at ultra-high vacuum for further experiments [30].

Injection into a Penning trap using collisions for friction requires some care. The ion's motion becomes instable since collisions lead to an increase of the magnetron radius. This can be overcome when the trap ring electrode is split into four segments and an additional radio-frequency field is applied between adjacent parts to create a quadrupolar rf field in the radial plane. At the sum of the perturbed cyclotron

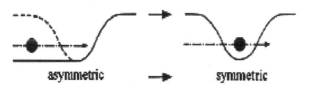

Fig. 17 Simple model describing the ion capture in a Penning trap

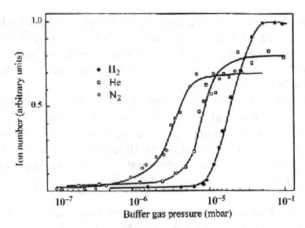

Fig. 18 Trapped Ba+ ion number vs. buffer gas pressure for He, H_2, and N_2 when injected at a few keV kinetic energy. The maximum ion number for H_2 is arbitrarily set to 1 (from [29])

frequency ω_+ and the magnetron frequency ω_- this field couples both oscillations. The damping of the cyclotron motion by collisions with the background atoms overcomes the increase of the magnetron radius, and as the result, the ions aggregate near the trap's center [23]. Since $\omega_+ + \omega_- = \omega_c$, the free ion cyclotron frequency, depends on the ion mass, it is possible to stabilize a particular isotope or even isobar in the trap by proper choice of the rf frequency [31] while ions of unwanted mass do not remain in the trap.

2.2 Trapped Particle Detection

Destructive Detection

(1) Paul Traps

Ions confined in a Paul trap may leave the trap either by lowering the potential of one endcap electrode or by application of a voltage pulse of high amplitude. They can be counted by suitable detectors such as ion multiplier tubes or channel plate detectors. By different arrival times to the detector, ions of different charge-to-mass ratio can be distinguished (Fig. 19). The amplitude of the detector pulse depends on the phase of the leading edge of the ejection pulse with respect to the phase of the rf trapping voltage.

When several ion species are simultaneously trapped, a particle of specific charge-to-mass ratio can be selectively ejected from the trap and counted by a detector when a rf field resonant with the axial oscillation frequency of the ion of interest is applied in a dipolar mode between the trap's endcap electrodes [33].

Fig. 19 Ejection of simultaneously trapped H^+, H_2^+, and
H_3^+ ions from a Paul trap and
detection by an ion multiplier
located 5 cm from the trap's
endcap (from [32])

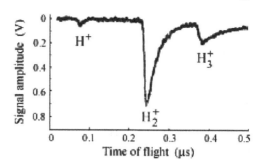

(2) Penning Traps

Ions can be released from the Penning trap, generally placed in the most homogeneous part of a (superconducting) solenoid, by switching one of the endcap potentials to zero. The particles then travel along the magnetic field lines until they
arrive at a detector outside the magnetic field region (Fig. 20). At finite radial kinetic energies, the ions have some angular momentum which is associated with
a magnetic moment μ. In the fringe field of the magnet, a force $F = \mathrm{grad}(\mu B)$
acts upon the ions and accelerates them onto the detector. The time of flight for a
given angular momentum is determined by the ions' mass and thus simultaneously
trapped different mass ions can be distinguished. We note that excitation of the radial motion by an additional rf field increases the angular momentum and thus leads
to a reduced time of flight. This method of detecting ion oscillation frequencies,
particularly the cyclotron frequency, serves as basis for many high-precision mass
spectrometers [34].

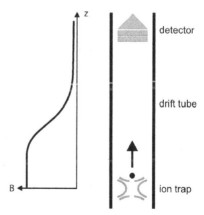

Fig. 20 Time-of-flight detection of ions released from a
Penning trap. The magnetic
field points in the vertical
direction

Non-destructive Detection

Tank Circuit Damping

The mass-dependent oscillation frequencies of ions in a trap can be used for detection without ion loss. A tank circuit consisting of an inductance L and the trap electrodes as capacitance C is applied to the trap and weakly excited by its resonance frequency ω_{LC} (Fig. 21). The ions' axial oscillation frequency ω_z can be changed by variation of the electric trapping field. When both frequencies coincide, energy is transferred from the circuit to the ions leading to a damping of the circuit. This results in a decrease in the voltage across the circuit. Modulation of the trap voltage around the operating point at which resonance occurs and rectification of the voltage across the circuit lead to a repeated voltage drop whose amplitude is proportional to the number of trapped ions. When different ions are simultaneously confined, signals appear at different values of the modulated trapping voltage. The sensitivity of the method depends on the quality factor Q of the resonance circuit. With moderate values of the order of $Q = 50$, about 1000 trapped ions lead to an observable signal.

Bolometric Detection

In the bolometric detection of trapped ions, first proposed and realized by Dehmelt and coworkers [35], the ions are kept in resonance with a tuned LC circuit connected to the trap electrodes in a way similar to what was discussed in the previous paragraph. An ion of charge q oscillating between the endcap electrodes of a trap induces a current in the external circuit given by

$$I = \Gamma \frac{q \dot{z}}{2z_0} , \tag{42}$$

where $2z_0$ is the separation of the endcaps and Γ is a correction factor, which accounts for the approximation of the trap electrodes by parallel plates of infinite dimension. For hyperbolically shaped electrodes, $\Gamma \approx 0.75$. The electromagnetic

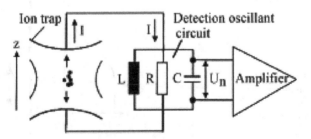

Fig. 21 Electronic circuit for non-destructive detection of trapped ion clouds

energy associated with this current will be dissipated as thermal energy in the parallel resonance resistance of the LC circuit, R. The increased temperature T of that resistance results in an increased thermal noise voltage U_{noise} per bandwidth δv:

$$U_{noise} = \sqrt{4kTR\delta v} \,. \tag{43}$$

If the number of stored ions is known, the noise voltage measured by a narrow band amplifier can serve as a measure of the ion temperature. When ions of different charge-to-mass ratios are stored simultaneously, their oscillation frequencies can be brought into resonance with the circuit by sweeping the trap voltage. Each time an ion species is resonant with the circuit, the noise amplitude increases and can be detected. An example is shown in Fig. 22 where the increased noise of the axial resonances of different highly charged ions in a Penning trap is recorded. For high sensitivity, the thermal noise power of the circuit has to be kept as low as possible. When superconducting circuits at temperatures around 4 K are used, single ions can be well observed.

Fourier Transform Detection

A Fourier transform of the noise induced by the oscillating ions in a tank circuit shows maxima at the ion oscillation frequencies. When superconducting circuits with quality factors Q of several thousands are used, the signal strength from a single ion is sufficient for detection. Figure 23 shows the noise induced by six ions in a circuit attached to two segments of the ring electrode of a Penning trap. The

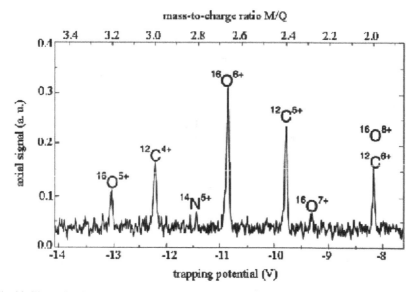

Fig. 22 Thermal noise induced in an LC circuit by simultaneously trapped ions of different charge-to-mass ratios in a Penning trap (from [36])

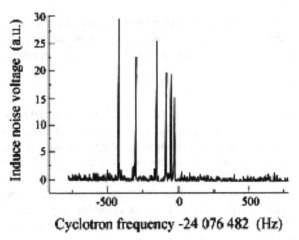

Fig. 23 Fourier transform of the noise from six C^{5+} ions induced in a tank circuit attached to two segments of the ring electrode of a Penning trap. The Q value of the circuit was 4000. The magnetic field was inhomogeneous and ions sitting at different places in the trap have slightly different cyclotron frequencies (from [36])

magnetic field was inhomogeneous, and ions sitting at different places in the trap have slightly different cyclotron frequencies.

When the ions are kept continuously in resonance with the tank circuit, the increased thermal energy of the detection circuit, due to the induced ion currents, will be dissipated. Consequently, the ion oscillation will be exponentially damped with a time constant τ given by

$$\tau = \frac{(2z_0)^2}{\Gamma^2 R} \frac{m}{q^2},\tag{44}$$

where R is the impedance of the circuit. When the ions reach thermal equilibrium with the circuit, excess noise can no longer be detected; nevertheless, the presence of ions can be detected by a spectral analysis of the circuit noise. The voltage that the ion induces with its remaining oscillation amplitude adds to the thermal noise of the circuit, however with opposite phase. As a result, the total noise voltage at the ion oscillation frequency is reduced, and the spectral distribution of the noise shows a minimum at this frequency when ions are present in the trap. This is shown in Fig. 24 for a single C^{5+} ion.

Optical Detection

A very efficient way to detect the presence of ions in the trap is to monitor their laser-induced fluorescence. This method is, of course, restricted to ions which have an energy level scheme which allows excitation by available lasers. It is based on the fact that the lifetime of an excited ionic energy level is of the order of 10^{-7} s when it decays by electric dipole radiation. Repetitive excitation of the same ion by a laser at saturation intensity then leads to a fluorescence count rate of 10^7 photons per

Fig. 24 Fourier transform of the noise of a superconducting axial resonance circuit in the presence of a single trapped C^{5+} ion in thermal equilibrium with the circuit. The sum of the circuit's thermal noise and the induced noise from the oscillating ion leads to a minimum at the ion's oscillation frequency

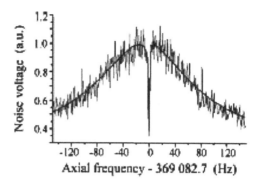

Fig. 25 Laser-induced fluorescence from 10 Ca^+ ions confined along the axis of a linear Paul trap taken with a CCD camera. Exposure time was 1 s. The distance between adjacent ions in about 15 μm (from [37])

second. Of those, a fraction of the order of 10^{-3} can be detected when we assume a solid angle of 10%, a photomultiplier detection efficiency of 10%, and filter and transmission losses of 90%, leading to an easily observable signal (Fig. 25).

The method is most effective when the ion under consideration has a large transition probability for the excitation from the electronic ground state as in alkali-like configurations. It becomes particularly easy when the excited state falls back directly into the ground state. Such two-level systems are available in Be^+ and Mg^+, and consequently, these ions are preferred subjects when optical detection of single stored particles is the issue. For other ions of alkali-like structure such as Ca^+, Sr^+, or Ba^+, it becomes slightly more complicated since the excited state may decay into a long-lived low-lying metastable state which prevents fast return of the ion into its ground state. Then an additional laser is required to pump the ion out of the metastable state. Signals of the expected strength have been obtained in all those ions. They allow even the visual observation of a single stored ion as demonstrated in a pioneering experiment at the University of Heidelberg on Ba^+ [38].

3 Ion Cooling Techniques

The mean kinetic energy of trapped ions depends on the operating conditions of the trap and the initial conditions of the ion motion. In thermal equilibrium, a typical value is 1/10 of the maximum potential depth. This is in general much higher than room temperature and might cause line shifts and broadening in high-precision spectroscopy. Therefore, a number of methods have been developed to reduce the temperature of the trapped particles.

3.1 Buffer Gas Cooling

Paul Traps

The influence of ion collisions with neutral background molecules can be treated by the introduction of a damping term in the equations of motion. Major and Dehmelt [39] have shown that ion cooling appears when the ions' mass is larger than that of the background molecule. The equilibrium ion temperature results from a balance between collisional cooling and energy gain from the rf trapping field. It depends on the buffer gas pressure (Fig. 26).

The thermalization process in a Paul trap concerns the macromotion of the ions while the amplitude of the micromotion remains fixed, determined by the amplitude of the electric trapping field at the ions' position. Cutler et al. [11] found that the calculated density distribution, averaged over the micromotion, changes from a Gaussian at high temperatures to a spherical one with constant density when the temperature approaches zero (see Fig. 7).

Penning Traps

Similar as in Paul traps, the axial motion of an ion in a Penning trap as well as the cyclotron motion is damped by collisions with neutral molecules. The magnetron oscillation, however, is a motion around an electric potential hill, and collisions tend to increase the radius of the magnetron orbit until the particle gets lost from

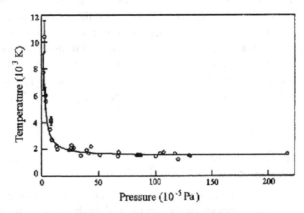

Fig. 26 Average kinetic energy of Ca^+ ions in a Paul trap at different N_2 background gas pressures [40]

the trap. This problem can be overcome by the introduction of an additional rf field in quadrupolar geometry at the sum frequency of magnetron and reduced cyclotron oscillations applied between four segments of the ring electrode. It couples both motions and the leads to an aggregation of the ions near the trap center [41, 42]. Figure 27 shows a simulation of the ion trajectory in the radial plane of a Penning trap under the influence of ion-neutral collisions as with and without the coupling field.

The initial radial amplitudes $\rho_+{}^0$ and $\rho_-{}^0$ of both motions change as

$$\rho_\pm(t) = \rho_\pm{}^0 \exp(\mp\alpha_\pm t) \tag{45}$$

and

$$\alpha_\pm = \frac{\gamma}{m} \frac{\omega_\pm}{\omega_+ - \omega_-} . \tag{46}$$

3.2 Resistive Cooling

When a LC circuit resonant with the ion oscillation frequency is attached between trap electrodes, the image currents induced by the ions' motion will lead to a current through the circuit and the increase in thermal energy will be dissipated to the environment [35, 43]. It leads to an exponential energy loss of the ions with a time constant

$$\tau = \left(\frac{2z_0}{q}\right)^2 \frac{m}{\omega L} \frac{1}{Q} , \tag{47}$$

where Q is the quality factor of the LC circuit. For an electron confined in a trap with $z_0 = 1\,\mathrm{cm}$, a circuit $Q = 100$ with $L = 1\,\mu\mathrm{H}$, oscillating at 300 MHz we obtain $\tau \approx 5\,\mathrm{ms}$. The increased time constant for heavier ions can be partly compensated by choosing superconducting circuits with Q values of several thousands. The final temperature will be the same as the temperature of the circuit which may be kept in contact with a liquid helium bath at 4 K.

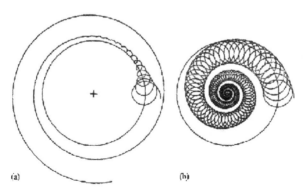

(a)   (b)

Fig. 27 Simulation of an ion radial trajectory in a Penning trap under the influence of collisions with neutral molecules without (**a**) and with (**b**) a quadrupolar rf coupling field at $\omega = \omega_+ + \omega_-$

Resistive cooling is particularly well suited for Penning traps. In Paul traps, the high amplitude of the rf trapping field will in general cause currents through the attached tank circuit which prevents reaching low temperatures.

We note that resistive cooling of an ion cloud affects the center-of-mass energy only. Coulomb interaction transfers energy from the individual ion oscillation into the center-of-mass mode. The rate of this energy transfer depends on the ion number. Thus, two time constants are observed (Fig. 28).

3.3 Laser Cooling

Laser cooling is the most effective way to reduce the ions' kinetic energy. It is based on the scattering of many photons from the same ion by repetitive excitation from the ground state to an excited state in a short time. Thus, the basic requirement is that the ion can be excited by electric dipole radiation and the excited state decays back into the ground level. This is the case in singly ionized Mg and Be ions which are most often used in laser cooling experiments. Other ions of alkali-like structure such as Ca^+, Sr^+, Ba^+, Hg^+, and others have more complex level schemes but can provide an effective two-level scheme by additional lasers.

Doppler Cooling

When a laser is tuned in frequency slightly below a resonance transition of an ion, only those ions absorb laser light with high probability which move against the direction of the laser beam because they are Doppler shifted into resonance. Photon recoil reduces their velocity. The absorbed photon is reemitted randomly in space

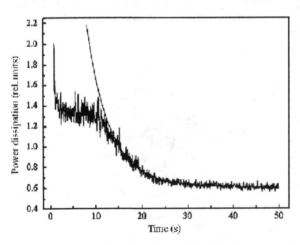

Fig. 28 Resistive cooling of an ion cloud in a Penning trap monitored by the induced image noise in an attached tank circuit. The center-of-mass energy is reduced in a short time of about 100 ms, while the individual ion oscillation is damped with a much longer time constant (5 s)

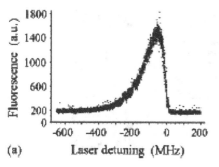 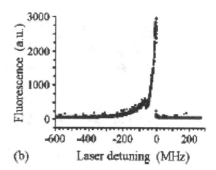

Fig. 29 Fluorescence from a small cloud of Ca^+ ions in a Paul trap when the laser frequency is slowly swept across the $4S_{1/2}$–$4P_{1/2}$ resonance. At laser frequencies below resonance ions are cooled, above resonance they are heated and the cloud expands. (**a**) Low laser power, (**b**) high laser power

and the average photon recoil momentum is zero. In order to cool stored ions, the laser frequency is swept from low frequency toward the resonance center. As soon as the laser frequency is higher than the ions' resonance, the opposite effect occurs: Ions moving in the direction of the laser beam absorb photons and gain energy. As a result, the ion cloud expands and the spatial overlap with the laser is reduced. When the fluorescence from the ion cloud is monitored, the resonance line shape becomes asymmetric showing a sharp edge at the high-frequency side (Fig. 29a).

The rate at which energy is lost depends on the laser detuning from resonance $\delta\omega$, the transition matrix element P, the ions mass m, and the laser intensity. The final temperature T_{min} is reached when the Doppler width of the transition is equal to the natural line width ("Doppler limit"):

$$T_{min} = \frac{\hbar\gamma}{2k_B} ,\tag{48}$$

where γ is the decay rate of the excited state and k_B the Boltzmann constant. Typical values for T_{min} are in the mK range.

When during the cooling process the temperature becomes so low, that the ratio Γ of the Coulomb repulsion energy between ions of average distance a and the thermal energy $1/2k_B T$

$$\Gamma = \frac{1}{4\pi\varepsilon_0} \frac{e^2}{ak_B T}\tag{49}$$

becomes larger than 175, a phase transition in the ion cloud to a crystalline structure appears. It manifests itself by a kink in the fluorescence line shape, caused by a sudden reduction of the Doppler broadening (Fig. 29b). The crystalline structures can be observed directly by a CCD camera (Fig. 30).

Sideband Cooling

When the amplitude of the ion oscillation in the harmonic trap potential becomes smaller than the wavelength of the exciting laser, no Doppler broadening appears.

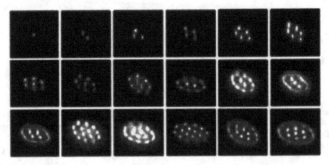

Fig. 30 Coulomb crystals of Ca$^+$ ions in a Paul trap observed by a CCD camera

Instead distinct sidebands in the excitation spectrum at multiples of the oscillation frequency show up ("Dicke effect"). The requirement is expressed by the condition $\eta \ll 1$, where η is the Dicke parameter:

$$\eta = k \left(\frac{\hbar}{2m\omega} \right)^{1/2} = \left(\frac{\omega_{rec}}{\omega} \right)^{1/2} . \tag{50}$$

Here k is the wave number of the laser radiation, ω the ion oscillation frequency, and $\omega_{rec} = (\hbar k^2)/2m$ the photon recoil energy. The sidebands are resolved when the spontaneous transmission rate γ is small compared to $\omega/2\pi$. This is in general only the case when transitions to long-lived metastable states are excited. An example is shown in Fig. 31.

If one irradiates the ion with a narrow band laser tuned to the first lower sideband at $\omega_0 - \omega_v$, where ω_0 is the resonance frequency of the ion at rest and ω_v the oscillation frequency, it absorbs photons of energy $\hbar(\omega_0 - \omega_v)$. The reemitted energy is

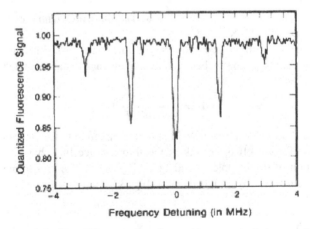

Fig. 31 Excitation of the $6S_{1/2}$–$5D_{5/2}$ quadrupole transition on a single laser-cooled ^{198}Hg$^+$ ion at 282 nm showing a carrier frequency at zero laser detuning from resonance and sidebands at the ions' micromotion frequency [44]

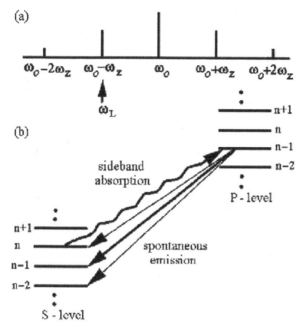

Fig. 32 Optical sideband cooling scheme

symmetrically distributed among carrier and sidebands, thus amounts on average to $\hbar\omega_0$ (Fig. 32). Hence, on the average, each scattered photon reduced the ions' vibrational energy by $\hbar\omega_v$ or, in a quantum mechanical picture, reduced the vibrational quantum number n by 1. Continuous excitation on the lower sideband finally drives the ion into the ground state of the confining potential. For a vibrational frequency of 1 MHz this corresponds to a temperature of $50\,\mu K$. It is indicated by the disappearance of the lower sideband in the excitation spectrum because for $n = 0$ no state with $n = -1$ exists in the excited optical level (Fig. 33).

Radiative Cooling

Accelerated charged particles loose energy by emission of radiation proportional to the square of their acceleration α:

$$\frac{dE}{dt} = \frac{-e^2}{6\pi\varepsilon_0 c^3}\alpha^2 . \tag{51}$$

For a particle oscillating with frequency ω, the mean energy loss is then given by

$$\frac{dE}{dt} = -\gamma E , \tag{52}$$

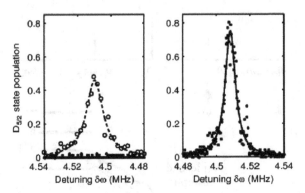

Fig. 33 Sideband cooling of a single Ca$^+$ ion into the vibrational ground state in a Paul trap: The lower and upper sidebands of the S$_{1/2}$–D$_{5/2}$ quadrupole transition after sideband cooling (*full circles*) are shown. The lower sideband after Doppler cooling only (*open circles*) is shown for comparison [45]

from which follows

$$E = E_0 e^{-\gamma t} , \tag{53}$$

with

$$\gamma = \frac{1}{6\pi\varepsilon_0} \frac{e^2 \omega^2}{mc^3} . \tag{54}$$

Because of the low mass and the high frequency, this energy loss is particularly significant for the cyclotron motion of electrons confined in a Penning trap at high magnetic fields while for atomic ions it can be totally neglected. In a field of 5 T, the time constant for electrons is calculated to $12\,\mathrm{s}^{-1}$. The final temperature is given by

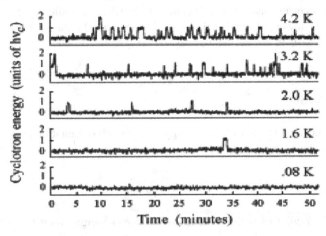

Fig. 34 Noise power induced in an electrode from a single electron confined in a Penning trap showing the occupation of the lowest quantum states of the cyclotron harmonic oscillator. At $T < 0.1\,\mathrm{K}$, the electron remains in the lowest quantum level for nearly infinite long times (from [46])

the equilibrium with the environment because of excitation by blackbody photons. At cryogenic temperatures, the quantum nature of the cyclotron oscillation becomes visible: The probability P_n of the population of a state with quantum number n is given by the Boltzmann distribution

$$P_n = A e^{-n\hbar\omega_c/k_B T} . \tag{55}$$

At $B = 5\,T$ and a temperature of $100\,mK$ the electron remains almost exclusively in the ground level ($n = 0$) of the cyclotron oscillation as experimentally demonstrated, Fig. 34 [46].

4 Summary

Charged particles can be confined by electromagnetic fields for practical unlimited time to a small volume in space. Efficient methods have been developed for loading of ions into the trap and for destructive or non-destructive detection. In the early period of trap ion physics, the low number of ions which can be stored simultaneously – limited by space charge to about 10^6 per cm^3 of active trap volume, several orders of magnitude less than the number of particles in neutral atom experiments – was considered a serious disadvantage. This was in part made up by the repetitive use of the same particles. In the course of time, however, the increased sensitivity of detection methods has allowed to observe even single particles with good signal-to-noise ratio. Then the earlier disadvantage turned into an advantage: A single isolated particle held in place by weak and well-controllable electromagnetic field, almost free of collisions when operated under ultra-high vacuum conditions, represents the best approximation of an ideal object for high-precision experiments.

In fact experiments of extremely high precision using single trapped ions have been performed in recent years in various fields of physics: Mass spectrometry has reached the level of 10^{-10} or below in fractional uncertainty [34], hyperfine structure separations and electronic g factors are determined with 10^{-12} and 10^{-9} uncertainty, respectively [36, 47], optical clocks based on single trapped ions have reached instabilities of a few parts in 10^{-16} [48], and, perhaps most remarkable, the g factor of the free electron has been determined to be below 10^{-12} [49, 50] providing the most accurate value of the fine structure constant α [51] and representing the most stringent test of quantum-electrodynamic calculations at low energies.

A potential disadvantage of ion traps is the fact that the kinetic energy of the stored ions is in general high compared to room temperature. It is determined by the initial trapping conditions and is typically of the order of a few eV. This would lead to large first-order Doppler broadening in optical transitions, and even second-order Doppler shifts might become significant. In recent years, however, efficient methods of ion cooling have been developed to reduce the ions' temperature to the quantum mechanical ground state of their oscillation. This not only leads to increased accuracy in spectroscopic experiments but allows precise control of the internal and external quantum states of trapped particles. It has opened a new field of physics, the potential application in quantum computation and quantum information [52, 53].

Acknowledgment I appreciate very much the help with the manuscript by Klaus Blaum and Joseba Alonso from the University of Mainz. The reported experiments from our laboratory were financially supported by the Deutsche Forschungsgemeinschaft, the German Ministry of Research and Technology, and the European Union.

References

1. F.G. Major, V. Gheorghe, and G. Werth, Charged Particle Traps, Springer, Heidelberg, (2005).
2. W. Paul, Rev. Mod. Phys. **62**, 531 (1990).
3. H. Dehmelt, Adv. At. Mol. Phys. **3**, 53 (1967).
4. J. Meixner and F.W. Schäffke, Mathieusche Funktionen und Sphäroidfunktionen, Springer, Heidelberg (1954).
5. N.W. McLachlan, Theory and Application of Mathieu Functions, Oxford University Press (1947).
6. R.F. Wuerker, H. Shelton, and R.V. Langmuir, J. Appl. Phys. **30**, 342 (1959).
7. X.Z. Chu et al., Int. J. Mass Spectrom. Ion Process. **173**, 107 (1998).
8. R. Iffländer and G. Werth, Metrologia **13**, 167 (1977).
9. F.G. Major and H.G. Dehmelt, Phys. Rev. **170**, 91 (1968).
10. H. Schaaf, U. Schmeling, and G. Werth, Appl. Phys. **25**, 249 (1981).
11. L.S. Cutler et al., Appl. Phys. B **36**, 137 (1985).
12. G. Kotowski, Z. Angew. Math. Mech. **23**, 213 (1943).
13. Y. Wang, F. Franzen, and K. Wanzek, Int. J. Mass Spectrom. **124**, 125 (1993).
14. R. Alheit et al., Int. J. Mass Spectrom. **154**, 155 (1996).
15. G. Tommaseo et al., Eur. Phys. J. D **28**, 29 (2004).
16. L.S. Brown and G. Gabrielse, Phys. Rev. A **25**, 2423 (1982).
17. M. Kretschmar, Z. Naturf. **45a**, 965 (1990).
18. L.S. Brown and G. Gabrielse, Rev. Mod. Phys. **58**, 233 (1986).
19. G. Bollen et al., J. Appl. Phys. **68**, 4355 (1990).
20. C. Gerz, D. Wilsdorf, and G. Werth, Nucl. Instrum. Meth. B **47**, 453 (1990).
21. R.S. Van Dyck et al., Phys. Rev. A **40**, 6308 (1898).
22. P. Paasche et al., Eur. Phys. J. D **22**, 183 (2003).
23. G. Savard et al., Phys. Lett. A**158**, 247 (1991).
24. Ch. Lichtenberg et al., Eur. Phys. J. D **2**, 29 (1998).
25. X.-P. Huang et al., Phys. Rev. Lett. **78**, 875 (1997).
26. E.M. Hollmann, F. Anderegg, and C.F. Driscoll, Phys. Plasmas **7**, 2776 (2000).
27. G. Bollen et al., Nucl. Instrum. Meth. A **368**, 675 (1996).
28. H.A. Schuessler and O. Chun-sing, Nucl. Instrum. Meth. **186**, 219 (1981).
29. J. Coutandin and G. Werth, Appl. Phys. B **29**, 89 (1982).
30. F. Herfurth et al., Nucl. Instrum. Meth. A **469**, 254 (2001).
31. H. Raimbault-Hartmann et al., Nucl. Instrum. Meth. B **126**, 378 (1997).
32. R. Alheit et al., Int. J. Mass Spectrom. Ion Process. **154**, 155 (1996).
33. F. Vedel and M. Vedel, Phys. Rev. A **41**, 2348 (1990).
34. K. Blaum, Phys. Rep. **425**, 1 (2006).
35. H.G. Dehmelt, Bull. Am. Phys. Soc. **7**, 470 (1962); H.G. Dehmelt and F.L. Walls, Phys. Rev. Lett. **21**, 127 (1968); D.A. Church and H.G. Dehmelt, J. Appl. Phys. **40**, 342 (1969).
36. G. Werth, H. Haeffner, and W. Quint, Adv. At. Mol. Opt. Phys. **48**, 191 (2002).

37. M. Block et al., J. Phys. B **33**, L375 (2000).
38. W. Neuhauser et al., Phys. Rev. A **22**, 1137 (1980).
39. F.G. Major and H.G. Dehmelt, Phys. Rev. **170**, 91 (1968).
40. F. Arbes et al., Z. Phys. D **25**, 295 (1993).
41. G. Bollen et al., J. Appl. Phys. **68**, 4355 (1990).
42. G. Savard et al., Phys. Lett. A **158**, 247 (1991).
43. D.J. Wineland and H.G. Dehmelt, J. Appl. Phys. **46**, 919 (1975).
44. J. Bergquist, W. Itano, and D.J. Wineland, Phys. Rev. A **36**, 428 (1987).
45. F. Schmidt-Kaler et al., J. Mod. Opt. **47**, 1573 (2000).
46. S. Peil and G. Gabrielse, Phys. Rev. Lett. **83**, 1287 (1999).
47. G. Werth, Phys. Scripta **T59**, 206 (1995).
48. S.A. Diddams et al., Science **306**, 1318 (2004).
49. R.S. Van Dyck, P.B. Schinberg, and H.G. Dehmelt, Phys. Rev. Lett. **59**, 26 (1987).
50. B. Odom, D. Hanneke, B. D'Urso, G. Gabrielse, Phys. Rev. Lett. **97**, 030801 (2006).
51. G. Gabrielse et al., Phys. Rev. Lett. **97**, 030802 (2006).
52. D. Kielpinski, C.R. Monroe, and D.J. Wineland, Nature **417**, 709 (2002).
53. H. Häffner et al., Nature **438**, 643 (2005).

Simulations for Ion Traps – Methods and Numerical Implementation

G. Zwicknagel

1 Introduction

The cooling of ions in traps and RF ion guides is a powerful tool for providing cooled ion bunches with low emittance and small energy spread which are, for example, needed for high-precision mass measurements at various rare-isotope facilities. This will be outlined in the associated article "Simulations for Ion Traps – Buffer Gas Cooling" by Stefan Schwarz, in which simulations of buffer gas cooling in ion traps and ion guides are discussed in detail. Such numerical simulations are indispensable for designing efficient cooling devices and for a theoretical guidance of related experiments. Here, particular questions to be addressed are, e.g., the optimization of the applied electric and magnetic fields, tests of proposed configurations and beam manipulations, performance optimizations, the explanation of observed unexpected behavior and the influence of space-charge effects.

This contribution deals with the basic concepts and the numerical realization of simulations which are commonly used for a theoretical description of various kinds of charged particle systems, like, e.g., ions in traps, ion beams or bunches of ions in accelerator structures or ions in ion sources. To this end the different approaches to the problem are introduced at first in Sect. 2: the direct particle–particle viewpoint taken in the molecular dynamics (MD) simulations and the phase-space picture underlying the frequently used particle-in-cell (PIC) scheme. In a second part, Sect. 3, we then outline the standard numerical techniques for calculating the forces, solving the Poisson equation and integrating the equations of motion which are needed for implementing these simulation methods. Concerning the integration of the equations of motion special emphasis is put in Sect. 3.4 on advanced, non-standard time integration schemes for specific applications like the ion motion in Penning traps or cases where the motion is dominated by a strong homogeneous magnetic field. The

G. Zwicknagel,
Institut für Theoretische Physik, Universität Erlangen-Nürnberg, Staudtstr. 7, 91058 Erlangen, Germany
e-mail: zwicknagel@theorie2.physik.uni-erlangen.de

Zwicknagel, G.: *Simulations for Ion Traps – Methods and Numerical Implementation*. Lect. Notes Phys. **749** , 69–96 (2008)
DOI 10.1007/978-3-540-77817-2_3

additional techniques used for buffer gas cooling like Monte Carlo simulations of the collisions of the ions with the buffer gas will be explained in the accompanying article by Stefan Schwarz and are omitted here.

2 Simulations of the Dynamics of Charged Particles

For the simulation of charged particle systems two different treatments are commonly applied: the direct particle–particle method realized in molecular dynamics (MD) simulations and the particle–mesh method implemented by the frequently used particle-in-cell (PIC) scheme, which is based on a phase-space picture of the charged particle system in terms of the Vlasov–Poisson equations. Both schemes have distinct regimes of application according to the amount of correlations which exist in the Coulomb systems at hand and have to be taken into account. Correlated systems are the domain of MD simulations, while PIC simulations are the method of choice for weakly correlated, collisionless systems where the space-charge, however, plays an important role.

Since all the charged particle systems we are interested in here, i.e., ions in traps, accelerator structures or ion sources, have rather low densities, they can be treated by classical Boltzmann statistics and by a classical dynamics of point-particles. Also radiation fields are negligible and the interaction between the charges is given by the related electrostatic fields. We thus restrict the following considerations to such kind of Coulomb systems and, for simplicity, further assume that they only consist of one species of point charges (ions). An extension of all the forthcoming expressions to multi-component systems is straightforward and can be easily undertaken.

2.1 Dynamics of Point Charges

We consider a system of N classical, point charges (q, m) under the influence of external electric $E_{\text{ext}}(r,t)$ and magnetic $B_{\text{ext}}(r,t)$ fields, which we suppose to be known. Usually they must be first calculated for a given specific application with appropriate numerical tools. The dynamics of such a classical N-particle system simply evolves according to Newton's law:

$$\frac{d}{dt}r_i = v_i \,, \tag{1}$$

$$m\frac{d}{dt}v_i = F_{ci} + q\left[E_{\text{ext}}(r_i,t) + v_i \times B_{\text{ext}}(r_i,t)\right] \,, \tag{2}$$

$$F_{ci} = -q\nabla_i\Phi_c(r_i,t) = -q\nabla_i\left[\frac{q}{4\pi\varepsilon_0}\sum_{j\neq i}^{N}\frac{1}{|r_i-r_j|}\right]$$

$$= \frac{q^2}{4\pi\varepsilon_0}\sum_{j\neq i}^{N}\frac{r_i-r_j}{|r_i-r_j|^3} \,, \tag{3}$$

where the force on the ith particle (2) consists of the external Lorentz-force and the Coulomb force $\boldsymbol{F}_{ci}$ from all other point charges given in terms of the electrostatic potential Φ_c which is generated by these charges.

A straightforward solution of the dynamics established by (1) and (2) can be achieved by a numerical integration of these $6N$ coupled (first-order ordinary) differential equations where the internal forces $-q\nabla\Phi_c$ are calculated directly from the actual positions $\{r_i\}$ via (3). Such kind of numerical simulations are known as molecular dynamics (MD) simulations (see, e.g., [1, 2, 3]), which represent a so-called particle–particle (PP) method.

In our situation MD simulations are a natural and first choice for small systems of a few hundred up to a few thousand particles. They are conceptually easy and fairly simple to implement and include the Coulomb interaction on a microscopic level. But such a treatment of the long-range Coulomb interaction implies a computational effort for calculating the forces $\{\boldsymbol{F}_{ci}\}$ which scales with N^2. This makes this type of simulations inapplicable for systems with a much larger number of particles, as it is usually the case in the applications mentioned in Sect. 1. We therefore need appropriate methods for an approximate treatment of the Coulomb interaction between the particles. This is a longstanding issue in plasma physics, in particular concerning the description of non-neutral, one-component plasmas. One usually starts by splitting the Coulomb interaction into two contributions which differ in range: The short-range part is responsible for short-range correlations, like the prevention of small interparticle distances due to the Coulomb repulsion, and for violent collisions with large momentum and energy transfer. The long-range contribution is responsible for collective effects, like screening and plasma oscillations. The long-range contribution and the collective effects prevail in weakly coupled collisionless plasmas, which are characterized by a small classical plasma parameter Γ, where

$$\Gamma = \frac{q^2}{4\pi\varepsilon_0 a k_B T}, \qquad \frac{1}{a} = \left(\frac{4\pi}{3}n\right)^{1/3}, \qquad (4)$$

for a plasma with density n and temperature T. The classical plasma parameter Γ basically represents the ratio between the interaction energy $q^2/4\pi\varepsilon_0 a$ between two charges separated by the mean distance a and their thermal energy $k_B T$. In weakly coupled plasmas ($\Gamma \ll 1$) the particle dynamics is dominated by the thermal motion $k_B T \gg q^2/4\pi\varepsilon_0 a$ and hard collisions are rare. This can be seen by regarding the collision rate ν for Coulomb collisions with more than $90°$ deflection, which is $\nu = n v_{th}\sigma_c(v_{th}) \propto n v_{th}^{-3} \propto \omega_p\Gamma^{3/2}$ (where v_{th} is the thermal velocity and ω_p the plasma frequency; see, e.g., [4]). Therefore weakly coupled plasmas, where $\nu/\omega_p \ll 1$, are also called collisionless plasmas. The different particle dynamics in weakly and strongly coupled plasmas are illustrated in Fig. 1. It shows the projection onto a plane of the trajectories of a few selected particles out of MD simulations of a one-component plasma with $\Gamma = 0.1$ (left part) and $\Gamma = 10$ (right part). The strongly curved trajectories (right part) in a strongly coupled plasma well demonstrate the dominance of the Coulomb interaction. For strongly coupled systems like in Fig. 1 (right), or for even much stronger coupled ones like ion crystals in traps, where the thermal motion becomes negligible and Γ takes values of a few hundred or more,

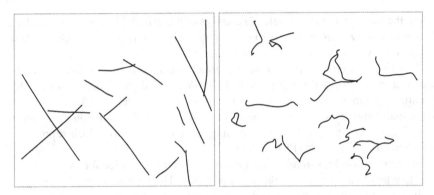

Fig. 1 Projection onto a plane of the trajectories of a few selected particles which have been obtained by following their positions for a certain time interval during MD simulations of plasmas of one species of charged particles with $\Gamma = 0.1$ (*left part*) and $\Gamma = 10$ (*right part*)

MD simulations, i.e., the numerical solution of (1)–(3), are the only appropriate treatment. In weakly coupled systems (Fig. 1, left) the thermal motion with straight trajectories clearly prevails. The Coulomb interaction mainly results in some scattering events with small deflection angles and some curvature of the trajectories in case where an overall macroscopic electric field (mean-field, space-charge field) is present. Weakly coupled systems thus allow for an approximative treatment by neglecting the short-range contributions of the Coulomb interaction. This strongly reduces the computational effort or permits, at the same expense, a much larger number of particles, as we will see later.

In most of the applications listed and discussed in the accompanying contribution "Simulations for Ion Traps – Buffer Gas Cooling," the charged particles are weakly coupled. If in addition the internal Coulomb interaction, i.e., the space-charge field $\nabla \Phi_c$, is small compared to the external field $(E_{ext} + v \times B_{ext})$ the full dynamics can be approximated by a pure single particle motion in this external field:

$$\frac{d}{dt}r_i = v_i, \quad m\frac{d}{dt}v_i = q\left(E_{ext}(r_i,t) + v_i \times B_{ext}(r_i,t)\right). \tag{5}$$

This requires, as in MD simulations, a numerical integration of these $6N$ differential equations by a proper time integration scheme, see Sect. 3.3. But since the equations of motion (5) for the single particle dynamics are not coupled via the Coulomb interaction (3) and only the given external fields are involved, the computational expense grows linear with the particle number N. Thus simulations with a few 10^6 particles can be easily performed on a PC and with many more particles on more powerful machines.

For weakly coupled systems ($\Gamma \ll 1$) where the space-charge field is not negligible compared to the external fields, a mean-field treatment can be employed. It neglects the short-range effects and correlations, but fully accounts for the long-range contributions and collective effects. This mean-field approach and the related PIC method will be discussed in detail in Sects. 2.2 and 2.3. For the moment we first give a simple picture of how a conceptually similar treatment can be achieved by just averaging out the short-range contributions. To this end we rewrite the microscopic space-charge field created by the ensemble of point charges, i.e., the electric potential

$$\Phi_c(r) = \frac{q}{4\pi\varepsilon_0} \sum_{j=1}^{N} \frac{1}{|r-r_j|} = \int \frac{d^3 r'}{4\pi\varepsilon_0} \frac{\rho(r')}{|r-r'|} \, , \tag{6}$$

in terms of the corresponding microscopic charge density

$$\rho(r) = \sum_{j=1}^{N} q \delta(r-r_j) \, . \tag{7}$$

Next we introduce a well-concentrated weight $w(r)$ function with a width σ of typically a couple of interparticle distances a [see (4)] and a smoothed charge density $\overline{\rho}$ defined by

$$\overline{\rho}(r) = \int d^3\tilde{r}\, w(r-\tilde{r})\rho(\tilde{r}) = \sum_{j=1}^{N} q w(r-r_j) \, . \tag{8}$$

From this smoothed charge density a smoothed electric potential with

$$\overline{\Phi}_c(r) = \int \frac{d^3 r'}{4\pi\varepsilon_0} \frac{\overline{\rho}(r')}{|r-r'|} \tag{9}$$

can be established. Both $\overline{\rho}$ and $\overline{\Phi}_c$ vary on a length scale σ, i.e., the width of the weight function w. Replacing F_{ci} in (2) with the "mean-field" $-q\nabla_i\overline{\Phi}_c$, and (3) with (9) results in

$$\frac{d}{dt}r_i = v_i \, ,$$
$$m\frac{d}{dt}v_i = -q\nabla_i\overline{\Phi}_c(r_i,t) + q\left(E_{ext}(r_i,t) + v_i \times B_{ext}(r_i,t)\right) \, , \tag{10}$$
$$\triangle\overline{\Phi}_c(r,t) = -\frac{\overline{\rho}(r,t)}{\varepsilon_0} \, , \tag{11}$$

where short-range contributions from the Coulomb interaction and correlations on a smaller scale than $\sigma \gg a$ are now absent. Instead of the straightforward particle–particle treatment (1)–(3), (10) and (11) are solved by a particle–mesh (PM) scheme. The particle motion (10) in the external fields and the mean-field $\nabla_i\overline{\Phi}_c$ is numerically integrated as in the case of the single particle motion (5). But the field $\overline{\Phi}_c$ is evaluated by solving the Poisson equation (11) for the smoothed potential $\overline{\Phi}_c$ on a grid. Here the mesh-size of the grid is $\sim\sigma$ and given by the length scale introduced by the averaged charge density $\overline{\rho}(r,t)$, which is sampled from the particle positions $\{r_j(t)\}$ according to (8). Details on solving the Poisson equation on a grid, which has to be done at least once per time-step, will be discussed in Sect. 3.1. The required numerical effort typically scales with the number of mesh-points to a power between one and two, depending on the chosen numerical implementation.

The computational effort for the particle motion (10) itself grows linearly with N and similarly large particle numbers can be treated as for the purely single particle motion (5). In the present scheme, however, the particle number also fixes the spatial extension of the whole grid and hence of the simulated volume: the number of grid cells with an edge length $\sim \sigma$ is limited by the number of particles in one cell which is $\propto (\sigma/a)^3 \gg 1$. On the other hand, the physically relevant length scale in weakly coupled plasmas is the Debye-length $\lambda_D = a/(3\Gamma)^{1/2}$ which sets the typical scale for collective effects and the screening of microscopic, short-range charge density fluctuations. Hence the mesh-size should be λ_D rather than σ which may require an unaffordable large number of particles for treating weakly coupled systems with $\Gamma \ll 1$. This is circumvented in the particle-in-cell (PIC) scheme by replacing the real particles with pseudo-particles of a larger mass and charge, retaining, however, the ratio q/m of the physical particles. But before discussing in Sect. 2.3 the PIC treatment in detail, we first turn to a statistical description of Coulomb systems in phase-space.

2.2 Dynamics in Phase-Space and Vlasov–Poisson Equations

Complementary to our previous consideration of point charges we now turn to a continuum approach by viewing our system of charged particles as a charged fluid. That is, we turn to a statistical approach where the positions and velocities of the point charges are replaced with the probability that a particle will be found with a given velocity at a certain location in space. Within this statistical continuum picture an ensemble of N charged particles, e.g., a cloud of ions in a trap or an ion beam in a storage ring, is described by the single particle density $f(r,v,t)$ in the 6D phase-space of position and velocity (or momentum). Here the probability $P(r,v,t)$ to find a particle at time t with velocity v at position r is given by

$$P(r,v,t) = \frac{1}{N}f(r,v,t)d^3r\,d^3v\,, \qquad \frac{1}{N}\int d^3r \int d^3v\, f(r,v,t) = 1\,. \qquad (12)$$

Accordingly, we can identify the particle density $n(r,t)$ as

$$n(r,t) = \int f(r,v,t)\, d^3v\,, \qquad \int n(r,t)\, d^3r = N \qquad (13)$$

and the velocity distribution $g(r,t)$ through

$$g(v,t) = \frac{1}{N}\int f(r,v,t)\, d^3r\,. \qquad (14)$$

Other quantities, e.g., the kinetic energy $E_{kin} = \int d^3r \int d^3v \frac{mv^2}{2} f(r,v,t)$, can be determined as expectation values with respect to $f(r,v,t)$. The time evolution of the system, i.e., the dynamics of point-particles as given by (1) and (2) in the particle picture, is now replaced by the dynamics of the density $f(r,v,t)$ in phase-space. It is determined by the Vlasov equation (see, e.g., [5] for its derivation and a detailed discussion)

$$\frac{\partial f}{\partial t} + v \cdot \frac{\partial f}{\partial r} + \frac{F}{m} \cdot \frac{\partial f}{\partial v} = 0 \,, \tag{15}$$

where $F(r,v,t)$ is the force acting on the particles. The single particle density $f(r,v,t)$ and its propagation according to (15) represent an exact description in the case of non-interacting particles where the force F is solely given by the external fields (as in (5)). For interacting particles it can be used as a reasonable approximation provided that particle correlations can be neglected, e.g., for weakly coupled, collisionless plasmas. In fact, such correlations have been discarded by reformulating the point-particle dynamics in terms of the single particle phase-space density and its time evolution. The single particle density contains, for instance, information about the probability to find a particle at a given position, but it does not carry any information, e.g., about the probability to find a certain distance between two particles, which is needed for determining the microscopic Coulomb field (3). The only source for the space-charge field $\Phi(r,t)$ we have at hand in the present picture is the charge density $\rho(r,t) = qn(r,t)$ as given through the particle density (13), where Φ follows from solving the Poisson equation $\triangle\Phi = -\rho/\varepsilon_0$. Together with the related force $-q\nabla\Phi$ and (15) this results in the Vlasov–Poisson equations:

$$\frac{\partial f}{\partial t} + v \cdot \frac{\partial f}{\partial r} = -\frac{q}{m}\left(-\nabla\Phi + E_{\text{ext}} + v \times B_{\text{ext}}\right) \cdot \frac{\partial f}{\partial v}, \tag{16}$$

$$\triangle\Phi(r,t) = -\frac{\rho(r,t)}{\varepsilon_0} = -\frac{q}{\varepsilon_0} \int f(r,v,t)\,d^3v \,, \tag{17}$$

where Φ must obey the existing boundary conditions for the electrostatic fields. Because two-particle (and higher-order) correlations are absent in f and thus in ρ, the dynamics of the system (16), i.e., the motion of the charged fluid, here evolves solely under the influence of the mean-field Φ as defined through ρ by the Poisson equation (17). The Vlasov–Poisson equations represent an appropriate description of a weakly coupled system of charged particles, in which the long-range collective contributions on a length scale of λ_D dominate over the short-range correlations. Compared with a pure single particle motion in external fields the presence of the mean-field Φ nevertheless strongly complicates the description because the Vlasov equation is now, via $\Phi = \Phi[f]$, non-linear in f, and (16) and (17) therefore require a self-consistent solution.

For practical applications of this mean-field description the Vlasov–Poisson equations must be solved numerically. The Poisson equation (17) can be treated, e.g., by finite differences on a 3D or, due to symmetries, on a lower dimensional grid or by an expansion in Eigen-functions of the Laplace operator. The basics of these procedures will be outlined in Sect. 3.2. Again the Vlasov equation (16) can be solved by representing f on a grid in phase-space and, e.g., using finite differences. This is a very demanding task and requires large computational resources as it has to be done on a 6D grid. An alternative and commonly used method is the PIC scheme, where the Vlasov equation is (approximatively) solved by means of a pseudo-particle dynamics.

2.3 The Particle-in-Cell (PIC) Scheme

The numerical solution of the Vlasov equation by the PIC technique [6, 7] starts by representing the phase-space density $f(r,v,t)$ (of N real particles) by a swarm of N_p pseudo-particles at positions $\{r_j\}$ and velocities $\{v_j\}$:

$$f(r,v,t) = \frac{N}{N_p} \sum_{i=1}^{N_p} w(r - r_i(t)) w_v(v - v_i(t)) . \tag{18}$$

The pseudo-particles have a mass $m_p = mN/N_p$ and a charge $q_p = qN/N_p$ which usually differ from the values for the physical particles, but the physical charge-to-mass ratio $q_p/m_p = q/m$ is the same. The w and w_v in (18) are smooth and well-concentrated, normalized weight functions in phase-space. As we will see below, only the spatial weight function w will explicitly enter the final numerical scheme and has to be chosen carefully, while the actual choice of w_v is of no relevance. For a time evolution of the phase-space density $f(r,v,t)$ of (18) according to the Vlasov equation (16), the pseudo-particles have to obey the equations of motion:

$$\frac{d}{dt}r_i = v_i , \qquad \frac{d}{dt}v_i = \frac{q_p}{m_p}\left(-\nabla\Phi(r_i,t) + E_{ext}(r_i,t) + v_i \times B_{ext}(r_i,t)\right) \tag{19}$$

in the external fields E_{ext}, B_{ext} and the space-charge field Φ. The space-charge field is produced from the charges of the pseudo-particles, i.e., via the Poisson equation (17) with the charge density

$$\rho(r,t) = q\frac{N}{N_p} \sum_{i=1}^{N_p} w(r - r_i(t)) . \tag{20}$$

The propagation of the Vlasov equation thus depends linearly on the number of pseudo-particles N_p. The PIC scheme is solved by a particle–mesh scheme where the equations of motion (19) are numerically integrated while the Poisson equation (17) (with ρ from (20)) is handled on a grid. The total spatial extension and shape of the grid and the number of grid points are regulated by the involved physical length scales (typically the Debye-length λ_D) and the dimensions of the system under consideration. The mesh size usually defines the width of the weight function w, and the number of pseudo-particles must be adapted to these settings in order to achieve a sufficiently smooth representation of the charge density on the grid. Typically, a few pseudo-particles per grid cell are sufficient. The use of a limited number of pseudo-particles results, however, in a certain amount of fluctuations or noise. It also introduces some effective two-particle interaction and collisions which are absent in the underlying Vlasov–Poisson description and may lead to artificial heating and damping. If necessary, such unwanted effects can be reduced by enhancing N_p and will vanish in the fluid limit $N_p \to \infty$, $q_p \to 0$, $m_p \to 0$ with $q_p/m_p = q/m$.

The PIC scheme (17), (19), (20) very much resembles (8), (10), (11), obtained by averaging the Coulomb interaction at short distances. This averaging suppresses short-range correlations, which are absent in the Vlasov–Poisson equations underlying the PIC treatment. But the previously outlined approach (8), (10), (11) is strongly limited by the use of physical particles. For a given computational effort and number of particles only a certain volume (of constant density) can be simulated. In the PIC treatment, on the other hand, the use of N_p pseudo-particles allows one to choose, within some limits, their charge q_p together with the width of the weight functions w in order to arrive at a description which properly accounts for the relevant physical length scales and the total size of the system.

3 Numerical Realization

From these basic ideas and concepts we now turn to their numerical realization. In both the particle–particle (MD) and the particle–mesh (PIC) scheme, the dynamics of the system is governed by sets of differential equations which first have to be approximated by algebraic equations required for numerical computations. This usually involves a discretization in time where the time evolution of the system then proceeds as a sequence of discrete (time)steps. In each step first the forces acting on the real or pseudo-particles must be calculated before an approximate integration of the equations of motion over this step can be performed. This is repeated over the required physical simulation time. Hence, the two central tasks are as follows:

1. The particle propagation by one time-step which is common to both simulation methods, PIC and MD, as well as to the single particle motion in external fields, (5).
2. The calculation of the force $F = -q\nabla\Phi$ from the (pseudo-)particle positions, either directly via (3) in the MD or by sampling ρ (20), solving the Poisson equation (17) and evaluating the gradient $\nabla\Phi$ on a grid in the PIC scheme. The handling of ρ, Φ, E_c on a grid is outlined in the next section, while an introduction to appropriate time integration schemes will be given in Sect. 3.3.

3.1 Implementation of the PIC (Particle–Mesh) Scheme

There are essentially two problems which have to be examined for the force calculation in a particle–mesh scheme. The first one is the relation between the pseudo-particles located at continuous positions $\{r_i\}$ and the involved quantities ρ, Φ, E_c existing only on the mesh-points $\{r_n\}$ of an appropriately chosen grid. For simplicity we assume an Cartesian grid with $r_n = (k\Delta_x, l\Delta_y, m\Delta_z)$, $k, l, m = 0, 1, 2, \ldots$.

This task is accomplished by assigning the pseudo-particle charge to the grid according to (20):

$$\rho(r_n) = q_p \sum_{i=1}^{N_p} w(r_n - r_i) \tag{21}$$

and by retrieving the forces on the particles through

$$F(r_i) = q_p \sum_n w(r_i - r_n)E(r_n) \quad \text{or} \quad F(r_i) = -q_p \sum_n [\nabla w(r_i - r_n)]\Phi(r_n). \quad (22)$$

Both the charge assignment (21) and the force evaluation (22) must be performed with the same weight function w in order to avoid a mismatch which would introduce spurious, unphysical forces on the particles. See, e.g., [6] for a detailed discussion of this effect. The weight function w itself together with the (affordable) number of pseudo-particles N_p and the grid-size (mesh-size and number of grid points) has to be chosen here as a good compromise between the total size of the system, the required spatial resolution (i.e., the physical length scale), the numerical noise and the computational expense. There exist various suitable weight functions which differ in shape and complexity of their implementation. Commonly used weight functions are, e.g., the rectangular profile (non-zero and constant over one mesh-size) or the triangular shape (non-zero over two mesh-sizes) in each dimension, see [6, 7] for more examples.

The second task is the solution of the Poisson equation $\triangle\Phi = -\rho/\varepsilon_0$, that is, the calculation of the potential $\Phi(r_n)$ from the known charge density $\rho(r_n)$ (21):

$$\Phi(r_n) = \sum_m G(r_n, r_m)\rho(r_m) \quad (23)$$

and the calculation of the electric field $E = -\nabla\Phi$:

$$E(r_n) = \sum_m D(r_n, r_m)\Phi(r_m) \quad (24)$$

on the grid. The functions G and D can be viewed as representations of the Green's function of the Laplace operator and the Nabla operator on the discrete grid, respectively. Examples for G and D are given in Sect. 3.2, where finite differences and Fourier transforms are briefly discussed as possible Poisson solvers. Figure 2 summarizes and illustrates one of the repeating time-step δt of the outlined PIC scheme, (21)–(24), together with a time-step in a pure particle–particle (MD) treatment (1)–(3). The particle propagation over one step, common to PIC and MD, will be discussed in Sect. 3.3.

3.2 Poisson Solver

To solve the Poisson equation (17) on a grid, i.e., for a finite set of values of ρ and Φ, various methods are in use; see, e.g., [6]. The actual choice strongly depends on the geometry and complexity of the system and the boundary conditions. Here we outline the basics of two possible approaches, finite differences and Fourier transforms, in a simple 3D Cartesian geometry.

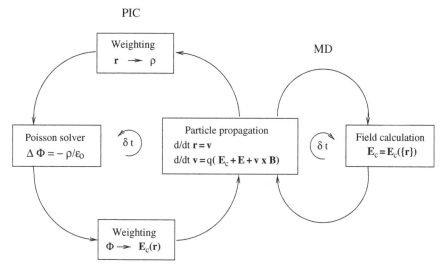

Fig. 2 Sketch of one time-step δt consisting of the field calculation and the particle propagation: on the *left* the particle–mesh PIC scheme, where the particles are propagated in a mean electric field, and on the *right* a particle–particle MD scheme, where the field is calculated as the sum of two-particle Coulomb interactions. E and B denote the externally applied fields and E_c the Coulomb field generated by the charged (pseudo-)particles

Finite Differences

The basic task is to find a representation of the Laplace operator and its inverse acting on the finite set of known values of Φ or ρ, respectively. Regarding for the moment only a one dimensional problem, we start with the question how to express the derivatives $h'_n = h'(x_n), h''_n = h''(x_n), \ldots$, of a continuous function $h(x)$ (with existing first- and second-order derivatives) at the grid points x_n, when $h(x)$ is known only at $\{x_n = n\Delta\}$ and represented by a finite set of values $\{h_n = h(x_n)\}$. To arrive at such expressions we perform a Taylor expansion about x_n with steps $\pm\Delta$:

$$h_{n+1} = h_n + h'_n \Delta + h''_n \frac{\Delta^2}{2} + h'''_n \frac{\Delta^3}{3!} + O(\Delta^4), \qquad (25)$$

$$h_{n-1} = h_n - h'_n \Delta + h''_n \frac{\Delta^2}{2} - h'''_n \frac{\Delta^3}{3!} + O(\Delta^4), \qquad (26)$$

where $h_n = h(x_n), h_{n\pm1} = h(x_{n\pm1} = x_n \pm \Delta)$ and $h_n^{(v)} = h^{(v)}(x_n)$.

Subtracting (26) from (25) and dividing the result by Δ yields the first-order derivative

$$h'_n = \frac{h_{n+1} - h_{n-1}}{2\Delta} + O(\Delta^2), \qquad (27)$$

with an error of order Δ^2, as the leading neglected term is here $h'''_n \Delta^2 / 3!$. In the same manner the second-order derivative

$$h_n'' = \frac{h_{n+1} - 2h_n + h_{n-1}}{\Delta^2} + O(\Delta^2) \tag{28}$$

is obtained by adding (25) and (26) and dividing by Δ^2. Since the leading neglected term is now $2h_n^{(4)}\Delta^2/4!$, the error of expression (28) is again $O(\Delta^2)$. Higher-order expression, i.e., with an error of $O(\Delta^3), O(\Delta^4), \ldots$, can be derived in a similar manner and are given, e.g., in [8]. They involve, however, more than the neighboring grid points. This strongly complicates, for instance, the implementation of boundary conditions. Thus one has to find some compromise between simplicity (and robustness), accuracy, mesh-size and numerical expense. Here second-order schemes are often a reasonable choice.

Going back to our original problem in three dimensions and taking (28), the Poisson equation $\triangle\Phi = -\rho/\varepsilon_0$ then reads

$$\begin{aligned}
\triangle\Phi &= \frac{\partial^2}{\partial x^2}\Phi + \frac{\partial^2}{\partial y^2}\Phi + \frac{\partial^2}{\partial z^2}\Phi \\
&= \frac{\Phi_{k+1,l,m} - 2\Phi_{k,l,m} + \Phi_{k-1,l,m}}{(\Delta_x)^2} + \frac{\Phi_{k,l+1,m} - 2\Phi_{k,l,m} + \Phi_{k,l-1,m}}{(\Delta_y)^2} \\
&\quad + \frac{\Phi_{k,l,m+1} - 2\Phi_{k,l,m} + \Phi_{k,l,m-1}}{(\Delta_z)^2} \\
&= -\frac{\rho_{k,l,m}}{\varepsilon_0}
\end{aligned} \tag{29}$$

with $\Phi_{k,l,m} = \Phi(k\Delta_x, l\Delta_y, m\Delta_z) = \Phi(r_n)$ and $\rho_{k,l,m} = \rho(k\Delta_x, l\Delta_y, m\Delta_z) = \rho(r_n)$.

This set of linear equations can also be written in matrix vector notation. As an example we consider a 1D case with $\triangle\Phi = \partial^2\Phi/\partial x^2 = -\rho/\varepsilon_0$, that is,

$$\Phi_{k+1} - 2\Phi_k + \Phi_{k-1} = -\frac{\rho_k}{\varepsilon_0}(\Delta_x)^2 := \tilde{\rho}_k , \tag{30}$$

and given boundary conditions for Φ_0, Φ_{N+1} at x_0, x_{N+1}. This can be expressed in the form

$$\underbrace{\begin{pmatrix} -2 & 1 & & & & \\ 1 & -2 & 1 & & & \\ & 1 & -2 & 1 & & \\ & & \cdot & \cdot & \cdot & \\ & & & \cdot & \cdot & \cdot \\ & & & 1 & -2 & 1 \\ & & & & 1 & -2 \end{pmatrix}}_{G^{-1}} \underbrace{\begin{pmatrix} \Phi_1 \\ \Phi_2 \\ \Phi_3 \\ \vdots \\ \Phi_{N-1} \\ \Phi_N \end{pmatrix}}_{\Phi} = \underbrace{\begin{pmatrix} \tilde{\rho}_1 - \Phi_0 \\ \tilde{\rho}_2 \\ \tilde{\rho}_3 \\ \vdots \\ \tilde{\rho}_{N-1} \\ \tilde{\rho}_N - \Phi_{N+1} \end{pmatrix}}_{\rho} \tag{31}$$

and solved for the electric potential $\Phi = G\rho$ by inverting G^{-1}. Here G can be identified (except for a factor Δ_x^2) as a representation of the Green's function G defined in (23). The matrix representation (31) can be extended straightforward to the 3D case by introducing appropriate vectors Φ and ρ, where some care is needed for properly defining the vector ρ as it now involves the more complex boundary conditions of the 3D case. By a proper arrangement of the components of the vectors $\boldsymbol{\Phi}$ and $\boldsymbol{\rho}$ the matrix G^{-1} itself will be composed of sub-matrices having the simple form of the matrix G^{-1} in (31). The final solution of the problem thus rests on the numerical inversion of sparse matrices for which very efficient standard algorithms have been developed. See, e.g., [9] and references therein for a discussion of available numerical tools.

In a similar manner as the Laplace operator via (28), the gradient can be expressed through (27), like for example $\boldsymbol{E} = -\nabla\Phi$ as

$$E_{k,l,m} = -\left(\frac{\Phi_{k+1,l,m} - \Phi_{k-1,l,m}}{2\Delta_x}, \frac{\Phi_{k,l+1,m} - \Phi_{k,l-1,m}}{2\Delta_y}, \frac{\Phi_{k,l,m+1} - \Phi_{k,l,m-1}}{2\Delta_z} \right) . \tag{32}$$

Rewriting this, again for 1D, in matrix vector notation gives

$$E = -\frac{1}{2\Delta_x} \underbrace{\begin{pmatrix} 0 & 1 & & & & \\ -1 & 0 & 1 & & & \\ & -1 & 0 & 1 & & \\ & & \cdot & \cdot & \cdot & \\ & & & \cdot & \cdot & \cdot \\ & & & -1 & 0 & 1 \\ & & & & -1 & 0 \end{pmatrix}}_{D} \underbrace{\begin{pmatrix} \Phi_1 \\ \Phi_2 \\ \Phi_3 \\ \vdots \\ \Phi_{N-1} \\ \Phi_N \end{pmatrix}}_{\Phi} + \frac{1}{2\Delta_x} \begin{pmatrix} -\Phi_0 \\ 0 \\ 0 \\ \vdots \\ 0 \\ \Phi_{N+1} \end{pmatrix} , \tag{33}$$

where $-D/2\Delta_x$ can be identified as a representation of the operator D introduced in (23). As before, the extension to 3D is obvious and only a question of bookkeeping.

Fourier Transform

Another way to solve the Poisson equation is the expansion of Φ and ρ in Eigenfunctions of the Laplace operator where the coefficients in these expansions are fixed by the finite set of values $\{\Phi(r_n), \rho(r_n)\}$ given at the grid points.

For a 3D Cartesian geometry an appropriate basis are plane waves $\exp(i\boldsymbol{k}\cdot\boldsymbol{r})$ with

$$\triangle \exp(i\boldsymbol{k}\cdot\boldsymbol{r}) = -k^2 \exp(i\boldsymbol{k}\cdot\boldsymbol{r}) . \tag{34}$$

This yields the Fourier representation of Φ and ρ

$$\Phi(r_n) = \sum_m \exp(i\boldsymbol{k}_m\cdot\boldsymbol{r}_n) \tilde{\Phi}(\boldsymbol{k}_m) , \qquad \rho(r_n) = \sum_m \exp(i\boldsymbol{k}_m\cdot\boldsymbol{r}_n) \tilde{\rho}(\boldsymbol{k}_m) , \tag{35}$$

with the discrete wave vectors $\{k_n\}$ of the reciprocal lattice and the Fourier coefficients

$$\tilde{\Phi}(\boldsymbol{k}_m) = \frac{1}{M} \sum_n \exp(-i\boldsymbol{k}_m\cdot\boldsymbol{r}_n) \Phi(r_n) ,$$

$$\tilde{\rho}(\boldsymbol{k}_m) = \frac{1}{M} \sum_n \exp(-i\boldsymbol{k}_m\cdot\boldsymbol{r}_n) \rho(r_n) . \tag{36}$$

Here $M = M_x M_y M_z$ is the total number of grid points (and allowed wave vectors).
By virtue of (34) the Poisson equation (17) is given in the Fourier domain by

$$-k_m^2 \tilde{\Phi}(k_m) = -\frac{\tilde{\rho}(k_m)}{\varepsilon_0} \implies \tilde{\Phi}(k_m) = \frac{\tilde{\rho}(k_m)}{k_m^2 \varepsilon_0}. \tag{37}$$

This now allows one to calculate $\Phi(r_n)$ from the known $\rho(r_n)$ by a forward Fourier
transform (36) of ρ and, after applying relation (37), a backward transform (35) of
$\tilde{\Phi}(k_m)$. This can be done very efficiently by using the fast Fourier transform (FFT)
technique; see, e.g., [9].

The Fourier transform procedure involving (35)–(37),

$$\Phi(r_n) = \frac{1}{M} \sum_m \sum_l \exp(ik_l \cdot r_n) \frac{1}{k_l^2 \varepsilon_0} \exp(-ik_l \cdot r_m) \rho(r_m)$$

$$= \sum_m G(r_n, r_m) \rho(r_m), \tag{38}$$

can also be reformulated, like in (31), in the matrix vector notation $\Phi = G\rho$ when
redefining $\Phi(r_n)$ and $\rho(r_n)$ as vectors Φ, ρ and the function $G(r_n, r_m)$ as a matrix G.

3.3 Time Integration Schemes

We now turn to the numerical treatment of the particle propagation, that is, the
numerical integration of the ordinary first-order differential equations:

$$\frac{d}{dt}r = v, \qquad \frac{d}{dt}v = \frac{1}{m}F(r, v, t). \tag{39}$$

For some general considerations and examples of possible integration schemes we
rewrite (39) in the more compact notation (in 6D phase-space):

$$x = (r, v); \quad f(x) = \left(v, \frac{F}{m}\right) \implies \frac{d}{dt}x = f(x). \tag{40}$$

One starting point for a numerical solution of the differential equation (40) is the
replacement of the continuous time variable t by a discrete set of time-levels $\{t_n\}$
separated by time-steps δt, i.e., $t \to \{t_n = n\delta t, n = 0, 1, 2, \dots\}$, and the introduction
of discrete sets of values: $x, f \to \{x_n = x(t_n), f_n = f(x_n)\}$. The exact integration
over one step $t_n \to t_{n+1} = t_n + \delta t$ which maps $x_n = x(t_n)$ onto $x_{n+1} = x(t_n + \delta t)$ is
given by

$$x_{n+1} = x_n + \int_{t_n}^{t_n + \delta t} dt' f\left(x(t')\right). \tag{41}$$

The task is now to find appropriate approximations for the time integral which only
involve the known values x_n and f_n. Quite a large variety of such approximations

and related integration schemes have been developed, see [9]. In the following we just present a view examples.

The simplest one is the so-called Euler step

$$x_{n+1} = x_n + \delta t f_n \,, \tag{42}$$

as obtained by simply taking $\int_{t_n}^{t_n+\delta t} f \, dt' = f_n \delta t$. The Euler step is a first-order $O(\delta t)$ scheme, that is, the error per step is proportional to $(\delta t)^2$. This can be easily seen, for instance, from the expansion $f(t) = f_n + f'_n(t - t_n) + O(\delta t^2)$ which yields for the integral in (41) $f_n \delta t + f'_n \delta t^2/2 + O(\delta t^3)$. Hence the error, given by the leading term neglected in (42), is $f'_n \delta t^2/2$. The Euler scheme is rather simple to implement and needs only one evaluation of f for one step δt. But it is rather inefficient and often unstable and one has to proceed for practical applications to more accurate higher-order approximations. This can be accomplished by using an Euler step (42) to first generate some intermediate values of x and $f(x)$ which then allow one to employ higher-order integration rules.

A second-order $O(\delta t^2)$ scheme can be derived using the so-called midpoint rule for evaluating the integral in (41) and reads

$$x_{n+1} = x_n + \delta t f_{n+\frac{1}{2}} \,, \qquad f_{n+\frac{1}{2}} = f\left(x_{n+\frac{1}{2}} = x_n + \frac{\delta t}{2} f_n\right) . \tag{43}$$

A similar second-order (predictor–corrector) scheme can be obtained by using $\tilde{x}_{n+1} = x_n + \delta t f_n$ as a first estimate for x_{n+1}. This provides an approximation for f_{n+1} by which we arrive at the final value for x_{n+1} using the trapezoidal rule for the time integral:

$$x_{n+1} = x_n + \frac{\delta t}{2} (f_n + f_{n+1}) \,, \qquad f_{n+1} = f(\tilde{x}_{n+1} = x_n + \delta t f_n) . \tag{44}$$

Both integration schemes, (43) and (44), are of second-order with an error proportional to $(\delta t)^3$ and need two evaluations of f per step.

An example of a fourth-order scheme, which is based on a higher-order integration rule and various estimates of intermediate values of x and f, is the (fourth-order) Runge–Kutta step with an error $\propto (\delta t)^5$ and four evaluations of f:

$$x_{n+1} = x_n + \frac{\delta t}{6} \left(f_n + 2 f_{n+\frac{1}{2}} + 2 \tilde{f}_{n+\frac{1}{2}} + \tilde{f}_{n+1}\right) \,, \tag{45}$$

in which

$$\tilde{f}_{n+\frac{1}{2}} = f\left(\tilde{x}_{n+\frac{1}{2}} = x_n + \frac{\delta t}{2} f_n\right) ,$$

$$f_{n+\frac{1}{2}} = f\left(x_{n+\frac{1}{2}} = x_n + \frac{\delta t}{2} \tilde{f}_{n+\frac{1}{2}}\right) ,$$

$$\tilde{f}_{n+1} = f\left(\tilde{x}_{n+1} = x_n + \delta t f_{n+\frac{1}{2}}\right) .$$

Important criteria for rating the various possible integration schemes are their consistency (e.g., time reversibility), accuracy, stability (error propagation) and the computational effort. The latter is usually dominated by the calculation of the force F, which requires, e.g., in the PIC treatment all the steps outlined in Sects. 3.1 and 3.2 including the necessary matrix operations or Fourier transforms. The computational effort needed for the propagation over a given time interval with a certain accuracy determines the efficiency of a scheme. Rather favorable in this respect are second-order schemes with only one force evaluation per step, like the leap-frog or the equivalent Velocity-Verlet [10] scheme (see (72), (73) later on). They can be applied in cases where the force $F = F(r,t)$ depends on the particle position r (and possibly also explicitly on the time) but not on the velocity, as it is usually the case when no magnetic field is present. The frequently used leap-frog scheme will thus be discussed in some detail next.

The Leap-Frog Scheme

The key feature of the leap-frog scheme is a shift of half a time-step between the time-levels where positions and velocities are given. We thus take r at times $t_n = n\delta t$ and v at $t_{n-\frac{1}{2}} = (n - \frac{1}{2})\delta t$ and reconsider (39). A Taylor expansion of $v(t)$ about $t_n, v_n = v(t_n)$ yields

$$v_{n-\frac{1}{2}} = v_n - \frac{F_n}{m}\frac{\delta t}{2} + \frac{1}{2m}\left(\frac{dF}{dt}\right)_n\left(\frac{\delta t}{2}\right)^2 + O(\delta t^3)\,, \tag{46}$$

$$v_{n+\frac{1}{2}} = v_n + \frac{F_n}{m}\frac{\delta t}{2} + \frac{1}{2m}\left(\frac{dF}{dt}\right)_n\left(\frac{\delta t}{2}\right)^2 + O(\delta t^3)\,. \tag{47}$$

By subtracting both equations the unknown velocity v_n and the unknown derivative of the force dF/dt cancel out and we have

$$v_{n+\frac{1}{2}} - v_{n-\frac{1}{2}} = \frac{F_n}{m}\delta t + O(\delta t^3)\,, \tag{48}$$

with $F_n = F(r_n, t_n)$. Similarly expanding $r(t)$ about $t_{n+\frac{1}{2}}$, $r_{n+\frac{1}{2}}$ results in

$$r_n = r_{n+\frac{1}{2}} - v_{n+\frac{1}{2}}\frac{\delta t}{2} + \frac{F_{n+\frac{1}{2}}}{2m}\left(\frac{\delta t}{2}\right)^2 + O(\delta t^3)\,, \tag{49}$$

$$r_{n+1} = r_{n+\frac{1}{2}} + v_{n+\frac{1}{2}}\frac{\delta t}{2} + \frac{F_{n+\frac{1}{2}}}{2m}\left(\frac{\delta t}{2}\right)^2 + O(\delta t^3)\,, \tag{50}$$

$$F_{n+\frac{1}{2}} = F\left(r_{n+\frac{1}{2}}, t_{n+\frac{1}{2}}\right) \tag{51}$$

and consequently

$$r_{n+1} - r_n = v_{n+\frac{1}{2}}\delta t + O(\delta t^3)\,, \tag{52}$$

Since $r_{n+\frac{1}{2}}$ and $F_{n+\frac{1}{2}}$ are canceled and hence all terms $O(\delta t^2)$. Combining (48) and (52) provides a second-order scheme (with an error of $O(\delta t^3)$) for the particle propagation by one step δt, i.e., $(r_n, v_{n-\frac{1}{2}}) \rightarrow (r_{n+1}, v_{n+\frac{1}{2}})$, with only one evaluation of F. It proceeds as follows: From the given r_n the force $F_n = F(r_n, t_n)$ is calculated, and the velocity is advanced from the given $v_{n-\frac{1}{2}}$ to the new

$$v_{n+\frac{1}{2}} = v_{n-\frac{1}{2}} + \frac{F_n}{m}\delta t \ . \tag{53}$$

With this new value for the velocity $v_{n+\frac{1}{2}}$ the new position

$$r_{n+1} = r_n + v_{n+\frac{1}{2}}\delta t \tag{54}$$

is obtained and one time-step is completed.

A Comparison of the Efficiency of Propagation Schemes

Figure 3 illustrates the efficiency of different propagation schemes for the case of a 1D harmonic oscillator with frequency ω [(58) with $F(t) = 0$]. It shows the averaged rms deviations of the position $x(t)$ predicted by the numerical integration schemes from the analytical solution (60). The average was taken over a total time of $100\,\omega^{-1}$. The rms deviation defines some measure for the error and thus for the efficiency of the used algorithm. It is given in the left part of Fig. 3 as a function of the size of the time-step δt and on the right as a function of the number of required force calculations. The performance of the different schemes is indeed as expected

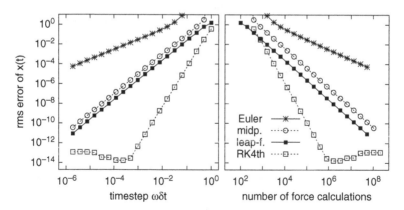

Fig. 3 Time averaged rms error of the position $x(t)$ of a 1D harmonic oscillator with frequency ω [(58) with $F(t) = 0$] as predicted by the numerical integration of the equations of motion over a total time of $100\,\omega^{-1}$ using different propagation schemes: Euler step (*stars*), (42); midpoint rule (*open circles*), (43); leap-frog scheme (*filled squares*), (53) and (54); and the fourth-order Runge–Kutta step (*open squares*), (45). On the *left* the error is given as a function of the size of the time-step, and on the *right* as a function of the total number of force calculations

from the estimated error in one time-step. For the first-order Euler step (stars), (42), the error diminishes $\propto \delta t$ for decreasing δt. As discussed in context with (42), the error in one step is here $\propto \delta t^2$. Since the total number of steps which are needed for a given time interval increases $\propto \delta t^{-1}$ for decreasing δt, the error with respect to the full time interval is, as observed, $\propto \delta t$. Similarly the error for the midpoint rule (open circles), (43), and the leap-frog scheme (filled squares), (53) and (54), is $\propto \delta t^2$, and is $\propto \delta t^4$ for the Runge–Kutta propagation (open squares), (45). For $\omega \delta t \gtrsim 0.01$ the Euler step is unstable and yields unrealistic results, while the error of the Runge–Kutta step stops to decrease for $\omega \delta t \lesssim 0.001$. At this point the error arrived at a value of about 10^{-14} and has reached the numerical accuracy (due to roundoff errors, etc.) of the machine on with these runs have been performed. For smaller time-steps the error even slightly increases as these errors accumulate with the increasing number of steps. For the present case of the harmonic oscillator the leap-frog procedure shows a somewhat better efficiency than the midpoint scheme, in particular with regard to the number of force calculations (right part of Fig. 3), i.e., with respect to the computational expense. Here the leap-frog propagation with one force evaluation per step is even competitive with the fourth-order Runge–Kutta scheme (with four force evaluations) if an accuracy of about 10^{-3} is still acceptable. The second-order predictor–corrector scheme (44) provides exactly the same results as the midpoint scheme because both procedures coincide for the harmonic oscillator.

The Leap-Frog Scheme for Particle Motion in a Magnetic Field

In many applications, as, e.g., in traps, an external magnetic field is present in addition to the electric field. In this case the force on the charged particles contains the velocity-dependent contribution $v \times B(r,t)$ and F_n also depends on v_n. This precludes the use of the leap-frog scheme and one has to employ, e.g., one of the second-order schemes like (43) and (44) with two force calculations per step. However, with some modification of the velocity propagation (53), the leap-frog algorithm can still be applied in the special, rather frequent case of a constant homogeneous magnetic field $B(r,t) = B$. Assuming a force $F(r,v,t) = qE(r,t) + qv \times B$ the velocity step (48) reads

$$v_{n+\frac{1}{2}} = v_{n-\frac{1}{2}} + \frac{q}{m} E_n \delta t + v_n \times \frac{qB}{m} \delta t + O(\delta t^3) . \tag{55}$$

Here the unknown value v_n can, by adding (46) and (47), be expressed through

$$v_n = \frac{v_{n+\frac{1}{2}} + v_{n-\frac{1}{2}}}{2} + O(\delta t^2) \tag{56}$$

and inserted into (55) (in a term which is already linear in δt). This yields, after some rearrangement,

$$v_{n+\frac{1}{2}} - \frac{v_{n+\frac{1}{2}}}{2} \times \frac{q\boldsymbol{B}}{m}\delta t = v_{n-\frac{1}{2}} + \frac{q}{m}\boldsymbol{E}_n\delta t + \frac{v_{n-\frac{1}{2}}}{2} \times \frac{q\boldsymbol{B}}{m}\delta t \; , \qquad (57)$$

with an error of $O(\delta t^3)$. The modified velocity propagation (57) can be cast in matrix vector notation and easily rewritten as an explicit expression for $v_{n+\frac{1}{2}}$ by an analytical inversion of the emerging matrices. The propagation of the position (54) remains unchanged since it only involves the (new) velocity $v_{n+\frac{1}{2}}$. This modified leap-frog scheme allows one to treat the particle motion in an electric field and a homogeneous magnetic field with only one evaluation of the force, i.e., of the electric field, in one step δt. Hence it is very efficient in cases where the electric field changes on a time scale of the order or shorter than the time scale given by the gyration of the charged particles in the magnetic field. But if the gyration sets the smallest time scale, as, e.g., for a strong magnetic field, most of the computing time is wasted for following the fast gyration and thus the known helical motion of the particles. Much more favorable under these specific conditions are time integration schemes which explicitly make use of known analytical solutions for that contribution to the motion which sets the shortest time scale. Some example of such schemes and their derivation are discussed in the next section.

3.4 Advanced Integration Schemes for Specific Applications

Before considering the motion in a strong magnetic field or a Penning trap, we first illustrate the idea behind these advanced, specific schemes and the conditions for their application for the case of a driven 1D harmonic oscillator, i.e., for the equation of motion:

$$\frac{d^2x}{dt^2} + \omega^2 x = \frac{F(t)}{m} \; , \qquad (58)$$

or, alternatively written as two first-order equations,

$$\frac{dx}{dt} = v \; , \qquad \frac{dv}{dt} = -\omega^2 x + \frac{F(t)}{m} \; . \qquad (59)$$

In the general case either the oscillator period, i.e., ω^{-1}, or the temporal variation of $F(t)$ will set the time scale to which the time-steps δt must be adapted for a numerical integration of (59) when using one of the previously discussed schemes. Thus for a driving force $F(t)$ which varies slowly with respect to the oscillator itself, i.e., $\left|\frac{1}{F}\frac{dF}{dt}\right| \ll \omega$, a time-step $\delta t \ll \omega^{-1}$ is required. On the other hand, we know (for $F = 0$) the analytical solution of the fast oscillator motion

$$x_{\text{osc}}(t) = x_0 \cos(\omega t) + \frac{v_0}{\omega}\sin(\omega t) \; , \qquad (60)$$

for the initial values $x_0 = x(t = 0), v_0 = v(t = 0)$. Moreover, the formal solution of the inhomogeneous equation (58)

$$x(t) = x_{\text{osc}}(t) + \int_0^t G(t-t') \frac{F(t')}{m} dt' \qquad (61)$$

can also be given in terms of (60) and the Green's function G of the harmonic oscillator

$$G(t-t') = \Theta(t-t') \frac{\sin(\omega(t-t'))}{\omega} , \qquad (62)$$

where $G(t-t')$ satisfies

$$\left(\frac{d^2}{dt^2} + \omega^2 \right) G(t-t') = \delta(t-t') , \qquad (63)$$

and $\Theta(t)$ is the step function. This yields the explicit expressions for $x(t)$ and $v(t)$:

$$x(t) = x_0 \cos(\omega t) + \frac{v_0}{\omega} \sin(\omega t) + \frac{1}{m\omega} \int_0^t \sin(\omega(t-t')) F(t') dt' , \qquad (64)$$

$$v(t) = -\omega x_0 \sin(\omega t) + v_0 \cos(\omega t) + \frac{1}{m} \int_0^t \cos(\omega(t-t')) F(t') dt' . \qquad (65)$$

They are appropriate starting points for a much more efficient numerical treatment of the driven oscillator because the remaining numerical integration of the time integrals over $F(t)$ can be done on a much larger time-step δt which is adapted to the slow variation of $F(t)$. For explicitly deriving such an integration scheme we consider one step, $t_n \rightarrow t_{n+1} = t_n + \delta t$, $(x_n, v_n) \rightarrow (x_{n+1}, v_{n+1})$, on this time scale and assume that $\omega \delta t \sim 1$. Using (64) we then get for the new position x_{n+1}:

$$x_{n+1} = x_n \cos(\omega \delta t) + \frac{v_n}{\omega} \sin(\omega \delta t) + \frac{1}{m\omega} \int_{t_n}^{t_n+\delta t} \sin(\omega(t_n + \delta t - t')) F(t') dt' . \qquad (66)$$

Since $F(t) = F(x(t), t) = F_n(x_n, t_n)$ is usually only known at t_n, the $F(t')$ must be suitably approximated. A possibility is the linear extrapolation $F(t') = F_n + F_n'(t' - t_n) + O(\delta t^2)$ with the derivative F_n' still undetermined. Ignoring for the moment F_n' and taking $F(t') = F_n + O(\delta t)$ the integral in (66) can be evaluated. This results, after some rearrangement, in

$$x_{n+1} = x_n \cos(\omega \delta t) + \frac{v_n}{\omega} \sin(\omega \delta t) + \frac{F_n}{m(\omega \delta t)^2} (1 - \cos(\omega \delta t)) \delta t^2 \; [+O(\delta t^3)] , \qquad (67)$$

which has an error $O(\delta t^3)$ [when counting $\omega \delta t$ as $O(1)$].

The derivation of the velocity step proceeds similarly, but is a bit more intricate. From (65) we have for the new velocity

$$v_{n+1} = -\omega x_n \sin(\omega \delta t) + v_n \cos(\omega \delta t) + \frac{1}{m} \int_{t_n}^{t_n+\delta t} \cos(\omega(t_n + \delta t - t')) F(t') dt' . \qquad (68)$$

To arrive at an error of $O(\delta t^3)$ as in the position step (67), we must now employ $F(t') = F_n + F_n'(t' - t_n) + O(\delta t^2)$. This yields

$$v_{n+1} = -\omega x_n \sin(\omega\delta t) + v_n \cos(\omega\delta t) \tag{69}$$
$$+ \frac{F_n}{m} \frac{\sin(\omega\delta t)}{\omega\delta t} \delta t + \frac{F_n'}{m} \frac{1 - \cos(\omega\delta t)}{(\omega\delta t)^2} \delta t^2 + O(\delta t^3) \,,$$

which, however, still requires an approximation for F_n'. For a force $F = F(x,t)$ which is independent of v, F_{n+1} can be calculated from the new position x_{n+1} obtained from (67) when doing that step first. With F_n and F_{n+1} at hand, F_n' can be represented by [cf. the case of spatial discretization (25)]

$$F_n' = \frac{F_{n+1} - F_n}{\delta t} + O(\delta t) \,. \tag{70}$$

Inserting this into (69) results in the final expression

$$v_{n+1} = -\omega x_n \sin(\omega\delta t) + v_n \cos(\omega\delta t) \tag{71}$$
$$+ \frac{F_n}{m} \frac{\sin(\omega\delta t)}{\omega\delta t} \delta t + \frac{F_{n+1} - F_n}{m} \frac{1 - \cos(\omega\delta t)}{(\omega\delta t)^2} \delta t \quad [+O(\delta t^3)] \,,$$

where the error is $O(\delta t^3)$ as for the advancement of x (67). Equations (67) and (71), subsequently executed in the proper order, thus provide a second-order integration scheme with only one force evaluation per step (as the new F_{n+1} is needed again in the next step) like in the leap-frog procedure. But here the time-step is fixed by the slow varying driving force and not by the fast oscillations.

Taking the present scheme, (67) and (71), in the limit $\omega \to 0$, when the driven oscillator, (59), reduces to the general equations of motion, (39) (here in 1D and with $F = F(x,t)$) yields [skipping error terms $\propto O(\delta t^3)$]

$$x_{n+1} = x_n + v_n \delta t + \frac{F_n}{2m} \delta t^2 \,, \tag{72}$$

$$v_{n+1} = v_n + \frac{F_{n+1} + F_n}{2m} \delta t \,. \tag{73}$$

This second-order scheme with only one force evaluation per step is the widely used Velocity-Verlet algorithm [10] which is equivalent to the leap-frog scheme, (53), (54). It has the advantage that position and velocity are known and stored at the same time-level t_n.

As an illustrative example we consider the harmonic oscillator (58) driven by the force

$$F(t) = \Theta(t)\Theta\left(\frac{\pi}{\Omega} - t\right) m\omega^2 x_0 \sin^4(\Omega t) \,, \tag{74}$$

which is non-zero only in the interval $0 < t < \pi/\Omega$ and where $\Omega \ll \omega$ and x_0 is the amplitude of the oscillator at $t = 0$. For the force (74) the analytical solution for the time evolution of the system can be easily calculated using the general expressions (64), (65) and then compared with the numerical integration with discrete time-steps δt. Such a comparison is made here with the previously discussed specialized integration scheme (67), (71) and the Velocity-Verlet algorithm (72), (73) but taking there as the force F the sum of the oscillator force $-m\omega^2 x$ and the driving force $F(t)$ (74). The resulting efficiency of these numerical schemes, monitored as in Fig. 3 by the rms deviation from the known exact position $x(t)$, is shown in Fig. 4 for $\Omega = 0.001\omega$ and an average over a total time-interval of $1.2\pi/\Omega$ starting at $t = 0$ with the initial conditions $x(t = 0) = x_0, v(t = 0) = 0$. The relevant time scale for the direct numerical integration by the Velocity-Verlet procedure is set by the smaller one of the characteristic times ω^{-1} and $\Omega^{-1} \gg \omega^{-1}$. And in fact the Verlet scheme reproduces the time evolution of the system with a reasonable accuracy only for time-steps well below $\omega\delta t = 0.1$ (see the open circles in Fig. 4). If the slowly varying driving force $F(t)$ (74) is set to zero all the time (stars in the center of the open circles), the same error shows up at all δt although the actual positions are different. With respect to the efficiency the (small) time-step needed to follow the fast oscillations automatically guarantees an accurate treatment of the slow evolving additional force $F(t)$. In contrast to that, a time-step of $0.1 \lesssim \omega\delta t \lesssim 1$ can be employed for an accurate numerical integration if the propagation scheme given by (67), (71) is applied (filled circles in Fig. 4). Interesting is here again the highest achievable accuracy, i.e., the lowest error which is in the present case for the Velocity-Verlet of about 10^{-6} at $\omega\delta t \approx 10^{-4}$ and for the specialized scheme (67), (71) of about 10^{-9} at $\omega\delta t \approx 0.03$. A further decrease of δt even results in an increasing error due to the increasing number of required steps and the related accumulation of roundoff errors. This can be nicely seen by the δt-dependence of the specialized scheme (67),

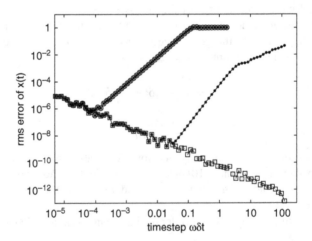

Fig. 4 Time-averaged rms error of the position $x(t)$ as in Fig. 3 but now for the numerical solution of the driven harmonic oscillator (58) with the driving force $F(t)$ given by (74). The time average is taken over the interval $0 \le t \le 1.2\pi/\Omega$. Shown are the observed errors as a function of the size of the time-steps resulting from a numerical integration using the specialized scheme given by (67), (71) (*filled circles*) and the Verlet algorithm, (72), (73), (*open circles*). For comparison the corresponding cases where $F(t)$ is set to zero over the whole time interval $0 \le t \le 1.2\pi/\Omega$ are also given (*open squares* and *stars*, respectively)

(71) when setting again the driving force to zero (open squares). Equations (67) and (71) then represent an exact solution for the propagation by one step δt. But their recursive evaluation up to the total time leads to an error which is continuously increasing with the number of steps, that is, with δt^{-1}. At sufficiently small time-steps the error of the Velocity-Verlet (open circles and stars) as well as of the specialized scheme (67), (71) with the full driving force (filled circles) falls onto this common curve (open squares) of achievable numerical accuracy set by the specifications of the machine on which the runs are performed.

Particle Propagation in a Strong Homogeneous Magnetic Field

In many applications the dynamics of charged particles proceeds under the influence of a strong external magnetic field which may dominate the motion and set the time scale. If in particular the motion of electrons is considered, their gyration in the magnetic field becomes the fastest process already at a moderate strength of the magnetic field. For ions this may happen, e.g., in the strong magnetic field in Penning traps, at a few Tesla.

As a further example where more efficient time integration schemes than, e.g., the modified leap-frog routine (54), (57) are applicable, we now discuss the propagation of charged particles in a constant homogeneous magnetic field $\boldsymbol{B} = B\boldsymbol{e}_z$ with the equations of motion

$$\frac{\mathrm{d}}{\mathrm{d}t}\boldsymbol{v} - \frac{q}{m}\boldsymbol{v} \times \boldsymbol{B} = \frac{\boldsymbol{F}(t)}{m}, \qquad \frac{\mathrm{d}}{\mathrm{d}t}\boldsymbol{r} = \boldsymbol{v} \tag{75}$$

or, rewritten for the components of $\boldsymbol{r}, \boldsymbol{v}$ and in terms of the cyclotron frequency $\omega_c = qB/m$,

$$\begin{pmatrix} \ddot{x} \\ \ddot{y} \end{pmatrix} - \omega_c \begin{pmatrix} 0 & 1 \\ -1 & 0 \end{pmatrix} \begin{pmatrix} \dot{x} \\ \dot{y} \end{pmatrix} = \frac{1}{m} \begin{pmatrix} F_x(t) \\ F_y(t) \end{pmatrix}, \qquad \begin{pmatrix} \dot{x} \\ \dot{y} \end{pmatrix} = \begin{pmatrix} v_x \\ v_y \end{pmatrix}, \tag{76}$$

$$\ddot{z} = \frac{1}{m} F_z(t), \qquad \dot{z} = v_z. \tag{77}$$

Here $\boldsymbol{F}(t) = (F_x, F_y, F_z)$ comprises all the other forces on the particle, i.e., the external electric field, the Coulomb field or space-charge field and generally also additional contributions related to an inhomogeneity of the magnetic field. We assume that this latter contribution is negligible, so that $\boldsymbol{F}(t) = \boldsymbol{F}(\boldsymbol{r}(t), t)$ is independent of $\boldsymbol{v}$, and that furthermore the force $\boldsymbol{F}(t)$ is slowly varying on the time scale of the gyration, $\left|\frac{1}{F_i}\frac{\mathrm{d}F_i}{\mathrm{d}t}\right| \ll \omega_c$. Under these conditions we can again make use of known analytical solutions, here for the helical motion of a charged particle in a constant magnetic field, in order to derive an integration scheme for a propagation with a time-step δt adapted to the variation of $\boldsymbol{F}(t)$, while $\omega_c \delta t \sim 1$. The motion along the magnetic field lines (77) is coupled to the transverse motion (76) via $\boldsymbol{F} = \boldsymbol{F}(x, y, z)$ only on the slow time scale δt. It can thus be handled efficiently by the

Velocity-Verlet routine, (72) and (73), and is not considered further. For the transverse motion the formal solution of (76) is

$$\begin{pmatrix} x(t) \\ y(t) \end{pmatrix} = \begin{pmatrix} x_0 \\ y_0 \end{pmatrix} + \frac{1}{\omega_c} A(t) \begin{pmatrix} v_{x,0} \\ v_{y,0} \end{pmatrix} + \frac{1}{m} \int_0^t G(t-t') \begin{pmatrix} F_x(t') \\ F_y(t') \end{pmatrix} dt' \quad (78)$$

for the initial conditions $x_0 = x(t = 0), y_0 = y(t = 0), v_{x,0} = v_x(t = 0), v_{y,0} = v_y$ $(t = 0)$. Here the (tensorial) Green's function G and the matrices A and B $\propto \dot{A}$ are defined by

$$G(t-t') = \frac{\Theta(t-t')}{\omega_c} A(t-t'), \quad A(t) = \begin{pmatrix} \sin(\omega_c t) & 1 - \cos(\omega_c t) \\ \cos(\omega_c t) - 1 & \sin(\omega_c t) \end{pmatrix}, \quad (79)$$

$$\frac{d}{dt} A(t) = \omega_c B(t), \quad B(t) = \begin{pmatrix} \cos(\omega_c t) & \sin(\omega_c t) \\ -\sin(\omega_c t) & \cos(\omega_c t) \end{pmatrix}. \quad (80)$$

Proceeding along the outline given in the previous example of the 1D oscillator we next consider (76) and its time derivative, i.e., $(v_x(t), v_y(t))$, for one step $t_n \rightarrow t_{n+1} = t_n + \delta t$ and $x_n, y_n, v_{x,n}, v_{y,n} \rightarrow x_{n+1}, y_{n+1}, v_{x,n+1}, v_{y,n+1}$. This yields

$$\begin{pmatrix} x_{n+1} \\ y_{n+1} \end{pmatrix} = \begin{pmatrix} x_n \\ y_n \end{pmatrix} + \frac{A(\delta t)}{\omega_c} \begin{pmatrix} v_{x,n} \\ v_{y,n} \end{pmatrix} + \int_{t_n}^{t_n+\delta t} \frac{A(t_n + \delta t - t')}{m\omega_c} \begin{pmatrix} F_x(t') \\ F_y(t') \end{pmatrix} dt', \quad (81)$$

$$\begin{pmatrix} v_{x,n+1} \\ v_{y,n+1} \end{pmatrix} = B(\delta t) \begin{pmatrix} v_{x,n} \\ v_{y,n} \end{pmatrix} + \frac{1}{m} \int_{t_n}^{t_n+\delta t} B(t_n + \delta t - t') \begin{pmatrix} F_x(t') \\ F_y(t') \end{pmatrix} dt'. \quad (82)$$

Inserting in (81), as in (66), $F(t') = F_n + O(\delta t)$ as an approximation to $F(t')$ provides after analytical integration the new transverse position

$$\begin{pmatrix} x_{n+1} \\ y_{n+1} \end{pmatrix} = \begin{pmatrix} x_n \\ y_n \end{pmatrix} + \frac{1}{\omega_c} A(\delta t) \begin{pmatrix} v_{x,n} \\ v_{y,n} \end{pmatrix} + \frac{1}{m(\omega_c \delta t)^2} C(\delta t) \begin{pmatrix} F_{x,n} \\ F_{y,n} \end{pmatrix} \delta t^2, \quad (83)$$

where

$$C(t) = \begin{pmatrix} 1 - \cos(\omega_c t) & -\sin(\omega_c t) + \omega_c t \\ \sin(\omega_c t) - \omega_c t & 1 - \cos(\omega_c t) \end{pmatrix}. \quad (84)$$

From the new x_{n+1}, y_{n+1} together with z_{n+1} from (72) one next calculates the updated force F_{n+1}, which again allows the use of the approximation $F(t') = F_n + [(F_{n+1} - F_n)/\delta t] (t' - t_n) + O(\delta t^2)$ in the velocity propagation (82). This yields the new velocity

$$\begin{pmatrix} v_{x,n+1} \\ v_{y,n+1} \end{pmatrix} = B(\delta t) \begin{pmatrix} v_{x,n} \\ v_{y,n} \end{pmatrix} \quad (85)$$

$$+ \frac{A(\delta t)}{m\omega_c \delta t} \begin{pmatrix} F_{x,n} \\ F_{y,n} \end{pmatrix} \delta t + \frac{1}{m(\omega_c \delta t)^2} C(\delta t) \begin{pmatrix} F_{x,n+1} - F_{x,n} \\ F_{y,n+1} - F_{y,n} \end{pmatrix} \delta t.$$

Equations (83), (85) together with the Velocity-Verlet, (72), (73), for the longitudinal motion, represent a second-order scheme with only one force evaluation per time-step and an error of $O(\delta t^3)$, where the time-step δt is assumed to be given by the temporal variation of the force $F(t)$, while $\omega_c \delta t \sim 1$. Like the previous scheme for the harmonic oscillator, (67), (71), the present scheme also turns into the Velocity-Verlet, (72), (73), now in the limit of vanishing magnetic field $\omega_c \to 0$.

The algorithm (83) and (85) for particle propagation in a homogeneous magnetic field was originally derived (in different ways) in [11]. It has been successfully used for numerical simulations of the energy transfer in collisions of ions and electrons in a strong magnetic field [12, 13, 14] and of the electron cooling of highly charged ions in Penning traps [15].

3.5 Motion of Charged Particles in a Penning Trap

For an efficient numerical treatment of the evolution of an ensemble of ions in a Penning trap, as it is needed, e.g., for a description of electron cooling of highly charged ions in the planned HITRAP facility [16, 17, 18], an even more specialized algorithm can be employed. Here the particles move in a strong homogeneous magnetic field, which provides the transverse trapping, and an electrostatic field composed of the longitudinally confining external electric field and the space-charge field. The motion in the magnetic field and the leading terms of the total electric field can be described by an analytical solution. The remaining forces are then treated as in the previous examples. In fact, the specific treatment for the particle propagation in a trap, which will be outlined next, is essentially a combination of the previous cases, i.e., the motion in a harmonic oscillator and the gyration in the magnetic field. Starting point is the motion in an ideal Penning trap, see, e.g., [19], where a single particle moves in a homogeneous magnetic field $\mathbf{B} = B\mathbf{e}_z$ and an (externally generated) electrostatic potential

$$\Phi = \frac{m}{2} \left[\kappa(x^2 + y^2) + \gamma z^2 \right] , \qquad (86)$$

with $2\kappa + \gamma = 0$ and $\gamma > 0$ for axial confinement. All deviations from this ideal case are for the moment put into the form of an additional (time-dependent) force $F(t)$. Hence the equations of motion are

$$\begin{pmatrix} \ddot{x} \\ \ddot{y} \end{pmatrix} + \kappa \begin{pmatrix} x \\ y \end{pmatrix} - \omega_c \begin{pmatrix} 0 & 1 \\ -1 & 0 \end{pmatrix} \begin{pmatrix} \dot{x} \\ \dot{y} \end{pmatrix} = \frac{1}{m} \begin{pmatrix} F_x(t) \\ F_y(t) \end{pmatrix} , \qquad \begin{pmatrix} \dot{x} \\ \dot{y} \end{pmatrix} = \begin{pmatrix} v_x \\ v_y \end{pmatrix} \quad (87)$$

$$\ddot{z} + \gamma z = \frac{1}{m} F_z(t) , \qquad\qquad \dot{z} = v_z , \qquad (88)$$

where $\omega_c = qB/m$ is again the cyclotron frequency. The numerical propagation of the axial motion (88) can be executed by the oscillator propagation routine, (67),

(71), and by identifying $\omega^2 = \omega_z^2 = \gamma > 0$. For the transverse motion the solution for the ideal case ($F = 0$) is well known and allows a formal solution of (87) which can be written, e.g., in the form

$$
\begin{pmatrix} x(t) \\ y(t) \end{pmatrix} = \frac{1}{\Delta\omega} [\omega_+ M_-(t) - \omega_- M_+(t)] \begin{pmatrix} x_0 \\ y_0 \end{pmatrix}
$$

$$
+ \frac{1}{\Delta\omega} \begin{pmatrix} 0 & 1 \\ -1 & 0 \end{pmatrix} [M_-(t) - M_+(t)] \begin{pmatrix} v_{x,0} \\ v_{y,0} \end{pmatrix} \tag{89}
$$

$$
+ \frac{1}{m\Delta\omega} \begin{pmatrix} 0 & 1 \\ -1 & 0 \end{pmatrix} \int_0^t [M_-(t-t') - M_+(t-t')] \begin{pmatrix} F_x(t') \\ F_y(t') \end{pmatrix} dt' ,
$$

for the initial values $x_0, y_0, v_{x,0}, v_{y,0}$ at $t = 0$. Here $\Delta\omega = \omega_+ - \omega_-$,

$$
M_\pm(t) = \begin{pmatrix} \cos(\omega_\pm t) & \sin(\omega_\pm t) \\ -\sin(\omega_\pm t) & \cos(\omega_\pm t) \end{pmatrix}, \quad \frac{d}{dt} M_\pm(t) = \omega_\pm \begin{pmatrix} 0 & 1 \\ -1 & 0 \end{pmatrix} M_\pm(t), \tag{90}
$$

and ω_+ and ω_- are the two Eigen frequencies of the ideal radial motion:

$$
\omega_\pm = \frac{\omega_c}{2} \left(1 \pm \sqrt{1 + \frac{4\kappa}{\omega_c^2}} \right). \tag{91}
$$

Here radial confinement is assumed, i.e., $\kappa = -\omega_+\omega_- > -\omega_c^2/4$ should hold for the given magnetic and electric fields.

Again one can benefit from the formal solution (89) in cases where $F(t)$ is slowly varying with respect to the fast motion given by ω_+. Then the time-step δt can be adapted to the time scale set by dF/dt or ω_-, while $\omega_+ \delta t \sim 1$, i.e., of $O(1)$. Completely along the previous lines of considerations, a second-order integration scheme with only one force evaluation per step can be derived from (89), if $F(t) = F(r(t),t)$, and reads for one step $t_n \to t_{n+1}$:

$$
\begin{pmatrix} x_{n+1} \\ y_{n+1} \end{pmatrix} = \frac{1}{\Delta\omega} [\omega_+ M_-(\delta t) - \omega_- M_+(\delta t)] \begin{pmatrix} x_n \\ y_n \end{pmatrix}
$$

$$
+ \frac{1}{\Delta\omega} \begin{pmatrix} 0 & 1 \\ -1 & 0 \end{pmatrix} [M_-(\delta t) - M_+(\delta t)] \begin{pmatrix} v_{x,n} \\ v_{y,n} \end{pmatrix} \tag{92}
$$

$$
+ \frac{1}{m\Delta\omega} \left(\frac{[M_-(\delta t) - 1]}{\omega_-} - \frac{[M_+(\delta t) - 1]}{\omega_+} \right) \begin{pmatrix} F_{x,n} \\ F_{y,n} \end{pmatrix},
$$

$$
\begin{pmatrix} v_{x,n+1} \\ v_{y,n+1} \end{pmatrix} = \frac{1}{\Delta\omega} \begin{pmatrix} 0 & 1 \\ -1 & 0 \end{pmatrix} [M_-(\delta t) - M_+(\delta t)] \left[\omega_+\omega_- \begin{pmatrix} x_n \\ y_n \end{pmatrix} + \frac{1}{m} \begin{pmatrix} F_{x,n} \\ F_{y,n} \end{pmatrix} \right]
$$

$$
+ \frac{1}{\Delta\omega} [\omega_+ M_+(\delta t) - \omega_- M_-(\delta t)] \begin{pmatrix} v_{x,n} \\ v_{y,n} \end{pmatrix} \tag{93}
$$

$$
+ \frac{1}{m\Delta\omega\delta t} \left(\frac{[M_-(\delta t) - 1]}{\omega_-} - \frac{[M_+(\delta t) - 1]}{\omega_+} \right) \begin{pmatrix} F_{x,n+1} - F_{x,n} \\ F_{y,n+1} - F_{y,n} \end{pmatrix}.
$$

Applications for the propagation scheme (92), (93) are for example simulations of the cooling of ions and antiprotons by electrons or positrons in Penning traps, as employed in recent experiments for the production of antihydrogen at CERN [20, 21] and the planned generation of slow highly charged ions at GSI [16, 17, 18]. In these nested traps, where an appropriate electric potential confines both the ions (or antiprotons) and a cloud of electrons (or positrons), the potential (86) of an ideal trap is, however, an improper starting point. The total electric potential $\Phi = \Phi_{ext} + \Phi_c$ composed of the given external field $\Phi_{ext}(r)$ and the space-charge field $\Phi_c(r,t)$ of the ions and electrons here strongly differs from the simple form (86) but may be piecewise fitted by the potential of an ideal trap. Dividing the trap in *a few* sections $[z_i, z_{i+1}]$ (along the axis, i.e., along the magnetic field lines), the potential Φ_i in each section is then given by

$$\Phi_i(r,t) = \Phi(z_i < z < z_{i+1}) = \frac{m}{2}\left[\kappa_i(t)(x^2 + y^2) + \gamma_i(t)z^2\right] + \delta\Phi_i(r,t), \qquad (94)$$

where we assume that $|\delta\Phi_i/\Phi_i| \ll 1$ and $\delta\Phi_i$, $\kappa_i(t)$, $\gamma_i(t)$ are slowly varying with respect to $\omega_z, \omega_\pm$ (91). This allows to use the scheme (92), (93) subsequently section by section with $F(t) = -q\nabla\delta\Phi_i(r,t)$, a constant $\kappa = \kappa_i$ and $\gamma = \gamma_i$, and a time-step adapted to the slow variations of $\delta\Phi_i, \kappa_i, \gamma_i$. For the piecewise fits of the potential of a real trap the axial motion may now involve solutions with a (piecewise) exponentially growing amplitude, i.e., $\gamma < 0$. In this case one has to set $\gamma = -\omega_z^2$ in the equation of motion (88) and replace ω_z with $i\omega_z$ and subsequently $\cos(i\omega_z t)$ with $\cosh(\omega_z t)$ and $\sin(i\omega_z t)$ with $i\sinh(\omega_z t)$ in the time integration scheme for the axial motion (67), (71). This outlined scheme has been recently employed for calculations of electron cooling of highly charged ions using the trap potential of the present HITRAP design, see [22] for details.

References

1. M.P. Allen and D.J. Tildesley, Computer Simulation of Liquids, Clarendon Press, Oxford (1987).
2. K. Binder and G. Cicctti (eds), Monte Carlo and Molecular Dynamics of Condensed Matter Systems, Società Italiana di Fisica, Bologna (1996).
3. D. Frenkel and B. Smit, Understanding Molecular Simulation: From Algorithms to Applications, 2nd edn, Academic Press, San Diego (2002).
4. S. Ichimaru, Basic Principles of Plasma Physics, 1st edn, Benjamin, Massachusett (1973).
5. R.L. Liboff, Kinetic Theory Prentice-Hall, New Jersey (1990).
6. R.W. Hockney and J.W. Eastwood, Computer Simulations Using Particles, McGraw-Hill, New York (1981).
7. C.K. Birdsall and A.B. Langdon, Plasma Physics via Computer Simulations, McGraw-Hill, New York (1985).
8. M. Abramowitz and I.A. Stegun, Handbook of Mathematical Functions, 10th edn, Dover, New York (1972).
9. W.H. Press, B.P. Flannery, S.A. Teukolsky, and W.T. Vetterling, Numerical Recipes, Cambridge University Press, Cambridge (1989).
10. L. Verlet, Phys. Rev. **159**, 98 (1967).

11. Q. Spreiter and M. Walter, J. Comput. Phys. **152**, 102 (1999).
12. G. Zwicknagel, Ion–electron collisions in a homogeneous magnetic field, In J.J. Bollinger, R.L. Spencer and R.C. Davidson, eds., AIP Conference Proceedings, vol. 498, AIP, New York (1999), pp. 469–474.
13. G. Zwicknagel and C. Toepffer, Energy loss of ions by collisions with magnetized electrons, In F. Anderegg, L. Schweikhard, and C.F. Driscoll eds., AIP Conference Proceedings, vol. 606, AIP, New York (2002), pp. 499–508.
14. H. Nersisyan, C. Toepffer, and G. Zwicknagel, Interactions Between Charged Particles in a Magnetic Field, Springer, Berlin Heidelberg New York (2007).
15. G. Zwicknagel, Electron cooling of highly charged ions in Penning traps. In M. Drewsen, U. Uggerhoj, and H. Knudsen, eds., AIP Conference Proceedings, vol. 862, AIP, New York (2006), pp. 281–291.
16. W. Quint et al., Hyperfine Int. **132**, 457 (2001).
17. Th. Beier et al., HITRAP, Technical Design Report, GSI Darmstadt (2003).
18. F. Herfurth et al., Int. J. Mass Spectrom. **251**, 266 (2006).
19. K. Blaum, Phys. Rep. **425**, 1 (2006).
20. M. Amoretti et al., Nature **419**, 456 (2002).
21. G. Gabrielse et al., Phys. Rev. Lett. **89**, 213401 (2002).
22. B. Möllers, Elektronenkühlung hochgeladener Ionen in Penningfallen, Ph.D. Thesis, University Erlangen, Erlangen (2007), http://www.opus.ub.uni-erlangen.de/opus/volltexte/2007/547/.

Simulations for Ion Traps – Buffer Gas Cooling

S. Schwarz

1 Introduction

Buffer gas cooling of ions in traps and radiofrequency ion guides has become the method of choice to improve the quality of continuous ion beams and to provide cooled ion bunches with low emittance and small energy spread. The method is chemically unselective and has consequently been applied to manipulate ion beams ranging from He to the heaviest elements. Depending on the buffer gas type and pressure, cooling times can be as low as a few milliseconds. The transmission efficiency of today's buffer gas coolers is approaching unity. These features make buffer gas coolers particularly useful at rare-isotope facilities, where speed, applicability, efficiency and good beam properties are essential to perform precision experiments with nuclei far from the valley of beta stability.

Two types of gas-filled coolers/bunchers are currently in use at rare-isotope facilities. They are based on linear Paul traps and on Penning traps: First the externally produced ions are electrostatically slowed down to a few electron-volt energy and injected into the gas-filled ion trap. For deceleration, the central trap electrodes are typically operated at a pedestal voltage close to beam potential. Inside the trap the ions are slowed down by collisions with buffer gas molecules and accumulate in the effective potential well provided by the ion trap. Finally, the cooled ion cloud is ejected out of the trap by a strong electric field and the ion pulse is then re-accelerated to a desired beam energy.

A radiofrequency (RF) multipole ion guide uses RF electric fields for the transverse confinement of ions and employs DC electric fields in the axial direction to

S. Schwarz

National Superconducting Cyclotron Laboratory, Michigan State University, East Lansing, MI 48824, USA

e-mail: schwarz@nscl.msu.edu

Schwarz, S.: *Simulations for Ion Traps – Buffer Gas Cooling.* Lect. Notes Phys. **749**, 97–117 (2008)

DOI 10.1007/978-3-540-77817-2_4

pull the ions through the buffer gas.[1] If the axial DC potentials are chosen such that they form a potential well to accumulate ions, then this device is usually referred to as a linear Paul trap. By switching the DC voltages the ions can then be extracted as a cooled ion bunch. This type of ion accumulator has become a workhorse at many rare-isotope facilities (see, e.g., [1, 2, 3, 4]) due to its high efficiency and its ability to deliver excellent pulses and has been key to a number of successful high-precision Penning trap mass measurements [5, 6, 7]. The application of RF buffer gas coolers in this field is steadily expanding. They are now also used to increase the sensitivity of collinear laser spectroscopy [8] and they are planned to improve the resolving power of mass separators at existing and future ISOL-facilities [9, 10].

Earlier approaches to accumulate ions at a rare-isotope facility were made at ISOLDE/CERN [11] with a 3D Paul trap [12], which uses RF electric fields in all three dimensions to confine ions. An improved version of this 3D Paul trap, installed at the ISOLTRAP mass spectrometer at ISOLDE/CERN, allowed for the first high-precision mass measurements of neutron-deficient mercury isotopes [13]. Despite its early success the 3D Paul trap was later replaced by a linear Paul trap, which provides higher efficiency, better beam properties and eases operation.

The second class of ion bunchers uses a gas-filled Penning trap. Such a system had first been applied as a cooler and isobar separator for the ISOLTRAP Penning trap mass spectrometer [14]; similar devices are now in use at the mass spectrometers CPT [15], SHIPTRAP [16] and JYFLTRAP [17] and for in-trap decay studies [18]. Because of its ability to handle rather large amounts of charges and to mass-selectively accumulate ions this type of cooler was chosen for the first beam preparation stage REXTRAP [19] of the REX-ISOLDE post-accelerator [20].

In order to arrive at the high performance of today's ion beam coolers, detailed simulations have proven critical. The simulations need to include the electric and/or magnetic fields present in these devices and to realistically model the buffer gas cooling process. The electromagnetic forces acting on the ions in ion bunchers can readily be obtained from Laplace solvers like Poisson Superfish [21] or SIMION [22] once the electrode layout has been defined. The established methods used to describe buffer gas cooling can be divided in a macroscopic and a microscopic approach. The macroscopic approach, discussed in the next section, provides a simple prescription of the time-averaged cooling force. The more realistic microscopic method tracks the cooling process by simulating individual encounters between the ion and the buffer gas molecule. This approach and its consequences for the ion motion are detailed in Sect. 3 and later.

The simulations presented in the following apply a number of simulation techniques reviewed in the accompanying article by Günter Zwicknagel, "Simulations for Ion Traps – Methods and Numerical Implementation." The reader is referred to this article for a more general discussion of simulation methods and details on their numerical realization.

---

[1] "Axial" refers to the axis of symmetry of the device, e.g., parallel to the rods of an ion guide. "Radial" refers to the two coordinates perpendicular to that axis.

2 Damping and Mobility, Macroscopic Approach

The time-averaged cooling effect of the buffer gas is often approximated by a viscous drag force. While this model does not take into account the statistical nature of the collisions of ions with the buffer gas molecules (which leads to model deficiencies that will be discussed later), it allows one to easily evaluate cooling times or assess the required dimensions of an ion trap under design. The damping force F_d, proportional to the velocity dr/dt of the ion with mass m, can be represented by a damping constant γ:

$$F_d = -\gamma m \frac{dr}{dt} .$$ (1)

For both Penning and Paul-type traps and reasonable experimental conditions this damping force results in an exponential damping of the amplitudes of motion (positions) proportional to $\exp(-\gamma t/2)$.[2]

Realistic damping constants γ for combinations of ions and buffer gases can be extracted from mobility data. A mobility measurement yields the drift velocity v_d with which an ion moves in a gaseous environment in the direction of an applied electric field E_{el}. The relationship between drift velocity and the strength of the electric field is characterized by the mobility K of the ion:

$$v_d = K \cdot E_{el} .$$ (2)

From the condition that the electric force balances the average damping force, the damping constant γ can be expressed by the mobility:

$$\gamma = \frac{q}{m} \cdot \frac{1}{K} .$$ (3)

As the mobility K scales with the pressure p and with temperature T, these dependencies are usually divided out and the reduced mobility is given as K_0:

$$K_0 = K \cdot \frac{273.16\,K}{T} \cdot \frac{p}{1013\,mbar} .$$ (4)

As long as the kinetic energy of the ion does not exceed a few electron-volts, the mobility is usually constant within a few percent. Figure 1 (left) shows literature mobility data for selected ions in noble gases at small drift velocities [23, 24, 25, 26] as a function of the mass of the buffer gas atom. Figure 1 (right) shows the cooling time $1/\gamma$, calculated from the mobility data according to (3), for a pressure of $p_{gas} = 10^{-4}$ mbar. Ion coolers and bunchers typically operate in the pressure range of $p_{gas} = 10^{-4}$–10^{-2} mbar, thus allowing for short cooling times in the sub-millisecond range.

---

[2] Assumptions: Paul trap: $\gamma \ll$ driving radiofrequency Ω; Penning trap: near-critically damped sideband cooling at the true cyclotron frequency. For details, see [27] and [28], respectively.

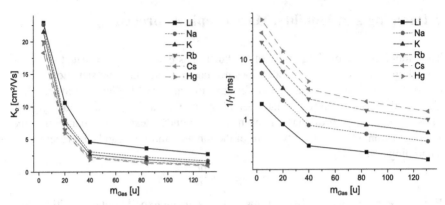

Fig. 1 *Left*: Mobility of selected ions in buffer gas (He, Ne, Ar, Kr, Xe) as a function of mass of the buffer gas atom. Data from [23, 24, 25, 26]. *Right*: Cooling time $1/\gamma$ calculated from the mobility data for a buffer gas pressure of $p_{gas} = 10^{-4}$ mbar

3 The Microscopic Picture

As Fig. 1 shows and data on other ion/gas combinations confirm, the mobility generally decreases for like ions as the mass of the buffer gas increases. This trend can already be expected from the simple theory of hard-sphere collisions, as a heavier buffer gas atom or molecule can carry away more of the ion's momentum in a collision. The picture of a more violent encounter of an ion with a heavier gas molecule remains valid if one replaces the hard-core potential used in hard-sphere collisions with more realistic potentials. Figure 2 shows trajectories of Cs ions and buffer gas atoms undergoing individual collisions with a He atom (a) and a Kr atom (b) using realistic interaction potentials. Under the conditions chosen for the two encounters, the Cs ion loses about 2% of its energy to the helium atom, but 67% to the krypton atom. If the cesium ion is placed in a weak uniform electric field in addition to the He or Xe buffer gas at a pressure of 10^{-3} mbar, the trajectories evolve as shown in Fig. 2c. Compared to inside He, the Cs ion takes a much more circuitous route inside the Xe gas to drift the required 1 m along the electric field; thus its drift time is higher and its mobility lower.

3.1 Necessity for a Microscopic Description, RF Heating

As shown above, the viscous drag model allows one to readily predict cooling times. This information, together with the energy to be dissipated, can then be used to estimate the required dimensions of an ion cooler, for example. However, the viscous drag model does not take into account the statistical nature of the collisions of ions with the buffer gas molecules and can significantly overestimate the effectiveness of the cooling process.

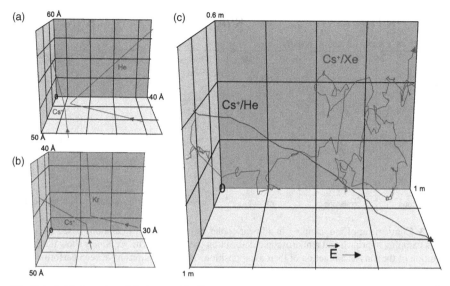

Fig. 2 *Left*: Trajectories of Cs ions and buffer gas atoms undergoing individual collisions. (**a**) Cs^+ with He; (**b**) Cs^+ with Kr. *Right*: Drift simulation of Cs ions in He and Xe buffer gas in the presence of a weak electric field

Figure 3 illustrates the model differences by comparing calculated trajectories based on a viscous damping approach and on a microscopic description. The calculation shows the cooling of K^+ ions in He, Ar and Kr buffer gas in a linear radiofrequency quadrupole ion cooler. The initial cooling rates are comparable for the two models, but then differences become apparent: The viscous drag calculations predict the confined ions to cool down to a complete standstill with zero temperature, while the collision model at best allows the ions to cool to the gas temperature. The microscopic calculations predict the cooling of K^+ in Ar to average out at a significantly increased temperature, the "cooling" of K^+ in Kr ends with the loss of the ion at one of the electrodes. Experimentally, these findings are confirmed by the fact that K^+ can be cooled in linear coolers with high efficiency in He, less efficiency in Ar and not at all in a Kr environment [1].

Ions stored in traps can get lost in collisions with gas molecules when the momentum gained in the encounter becomes too large for the trap to handle. Another, related loss mechanism present in RF-driven coolers (linear or 3D) is the so-called radiofrequency heating of ions [29], which becomes important when the mass of the buffer gas molecule M is comparable or larger than the mass of the ion m. This effect can be explained in a rather elegant way if the motion of the ion is viewed in the (three) phase-space areas spanned by the momentum and position components in each dimension, rather than in real space. In these phase-space areas, the trajectory of an ion is found on an ellipse for any given phase of the driving RF voltage. The only difference to the case of a simple harmonic motion is that this ellipse rotates

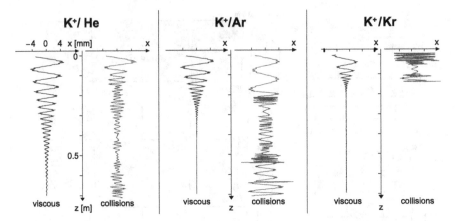

Fig. 3 Comparison of ion damping in a linear cooler as obtained by calculations based on a viscous damping approach and a microscopic description (collisions). The *six panels* show the radial position of the ions as a function of their axial position. The calculations have been performed for K^+ in He, Ar and Kr

as a rather involved (see next section) periodic function of the RF phase. To illustrate this fact, Fig. 4 shows trajectories for a weakly driven radial motion (Mathieu parameter $q_r = -0.02$) and for a strongly driven axial motion at $q_z = 0.8$ in phase space. Note that at times of identical RF phases $\xi = t\,\Omega/2$ the particles' paths touch the same ellipses.

Before illustrating the effect of RF heating itself let us first consider the collisional cooling process for a 1D harmonic electrostatic ion trap. In this type of ion trap (e.g., realized along the magnetic-field axis of a Penning trap or along the axis of symmetry in a linear Paul trap) the ion performs a simple harmonic motion. In phase space, the motion is represented by a closed ellipse, the area of which represents the total energy of the ion. In the event of a collision with a gas particle, the

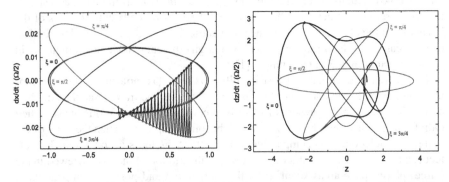

Fig. 4 Trajectories of ions in a Paul trap plotted in phase space. *Left*: A trajectory (*bold line*) for the radial motion, $q_r = -0.02$, as well as four ellipses for $\xi = 0, \pi/4, \pi/2$ and $3\pi/4$. *Right*: Idem, but for the axial motion with $q_z = 0.8$. For any given phase of the driving RF field the trajectories always touch the same ellipses

position of the ion does not change. If one assumes a gas particle at rest, the ion's momentum can change from near-zero to the negative of its initial value, depending on the ratio of the masses M/m. Independent of the amount of momentum transfer, the ion will be on an ellipse with a smaller area after the collision, i.e., the collisions will always lead to a cooling of the ion. For gas particles with a finite temperature, the cooling process is of course limited by exactly that temperature. This cooling process is illustrated in Fig. 5a and b. In an RF-driven trap, the situation is more complicated, since the ellipses in phase space rotate with time, as shown in Fig. 4 and has been discussed before. For simplicity, Fig. 5c shows only one particular ellipse, which the ion's path intersects at a certain RF phase. If there is only little momentum transferred, then the majority of the collisions lead to smaller elliptic paths, as illustrated in Fig. 5c. In this case there is an average cooling effect. For maximum momentum transfer, illustrated in Fig. 5d, the collisions lead to a larger elliptic path; in this case the ion picks up energy from the RF field. For $M = m$ the ion is at rest immediately after the collision, but then starts out on a larger orbit. For not-too-large RF amplitudes in a Paul-type trap the average energy transfer per hard-sphere collision can be calculated as a function of the mass ratio $\kappa = M/m$ [30]:

$$< \varepsilon_{RF} > = \kappa \, \frac{\kappa - 1}{(1 + \kappa)^2} . \tag{5}$$

For values $0 < \kappa < 1$ this expression is always negative, so a net cooling effect is expected. For $\kappa > 1$ the result becomes positive and in average more energy is absorbed from the RF field than dissipated in collisions. For equal masses there is no

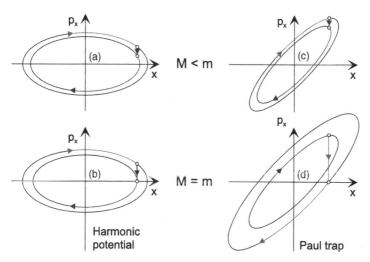

Fig. 5 Energy loss in collisions in traps. The two pairs of figures show simplified trajectories of an ion undergoing collisions in a 1D harmonic potential (**a,b**) and a Paul trap (**c,d**) in phase space. The pairs of cases (**a,c**) and (**b,d**) differ in the relationship of the masses of the ion m and the gas molecule M

Fig. 6 Average change in energy depending on the mass ratio of the collision partners M/m in a Paul trap and a harmonic potential

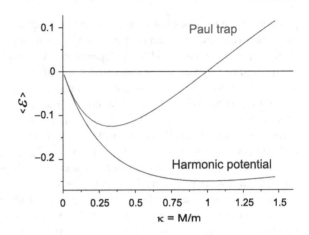

change in energy, as H. Dehmelt already argued for this case [31]. For comparison, the equivalent expression for the cooling in a static parabolic potential is

$$< \varepsilon_{par} >= -\frac{\kappa}{(1+\kappa)^2} \, . \tag{6}$$

The energy transfer in this case is always negative with a minimum when the two masses are equal. The dependencies (5) and (6) are shown in Fig. 6.

4 Realistic Interaction Potentials

Although the idea of collisions of hard spheres is quite useful to understand the cooling mechanism, one should not expect overly precise information on, e.g., cooling times from this model. In this simple model the interaction between the ion and the gaseous particles is only approximated reasonably well for higher energies of the ion. The long-range interaction of the ion with the buffer gas molecule (polarized by the ion) lets the gas act as a viscous medium, which lets the ion experience a damping force.

A class of potentials $V(r)$, which describe the interaction of an ion with a molecule at a distance r sufficiently realistically, are the so-called $(n,6,4)$ potentials [32]:

$$V(r) = \frac{B}{r^n} - \frac{C_6}{r^6} - \frac{C_4}{r^4} \, , \tag{7}$$

r being the distance between the ion and the buffer gas particle. The C_4/r^4 and C_6/r^6 terms quantify the attractive interaction of the ion's charge with the electric dipole and quadrupole moments, induced by the ion in the neutral particle. The short-range repulsive part of the interaction is described by the empirically determined term B/r^n. Equation (7) can alternatively be written as

$$V(r) = n\varepsilon \frac{\frac{12}{n}(1+\gamma_q)(\frac{r_m}{r})^n - 4\gamma_q(\frac{r_m}{r})^6 - 3(1-\gamma_q)(\frac{r_m}{r})^4}{n(3+\gamma_q) - 12(1+\gamma_q)}, \tag{8}$$

where r_m and ε are the position and depth of the minimum of the potential, respectively. γ_q is a measure of the strength of the r^6 term relative to the r^4 term.

The parameters in the ion-atom potentials can be derived by fitting the so-called collision integral, a theoretical expression for the collision cross-sections, to experimental data obtained from mobility data. The relationship between the collision integral and mobility will only be briefly sketched here; a thorough discussion of the theory can be found, e.g., in [32, 33]. The potential V and the mobility K can be connected by a series of integrations: The first one determines the scattering angle $\theta(b,E)$ as a function of the impact parameter b and the energy E available in the center-of-mass system:

$$\theta(b,E) = \pi - 2b \int_{r_a}^{\infty} \frac{1}{\sqrt{1 - \frac{b^2}{r^2} - \frac{V(r)}{E}}} \frac{dr}{r^2}. \tag{9}$$

The integration over the distance r between the ion and the gas molecule starts at the distance of closest approach of the collision partners r_a.[3] The collision integral (of first order) $\Omega(T)$ is obtained after two more integrations over the energy E and the impact parameter b. $\Omega(T)$ is a function of the temperature T:

$$\Omega(T) = \frac{4\pi}{(kT)^3} \int_0^{\infty} \exp\left(-\frac{E}{kT}\right) E^2 \int_0^{\infty} (1 - \cos(\theta(b,E))) b\,db\,dE, \tag{10}$$

k being the Boltzmann constant. $\Omega(T)$ is linked to the mobility K by

$$K(v_d) = \frac{3e}{16N} \sqrt{\frac{2\pi}{kT_{eff}(v_d)} \frac{m+M}{mM}} \frac{1}{\Omega(T_{eff}(v_d))}, \tag{11}$$

N being the number density of the gas molecules and T_{eff} an effective temperature, which accounts for both the gas temperature T_{gas} and the drift velocity v_d of the ion:

$$T_{eff} = T_{gas} + \frac{Mv_d^2}{3k}. \tag{12}$$

By successively inserting (7) or (8) into (9) and the result into (10) one obtains a relationship between the sets of potential parameters (n, B, C_6, C_4) or alternatively $(n, r_m, \varepsilon, \gamma_q)$ and the collision integral. The parameter C_4 is often known from the dipole polarizability of the gas particle. Information on the C_6 term is generally more scarce, but can be deduced from charge-induced quadrupole interaction energies and the $1/r^6$ term of the dispersion-type interaction energy (see [32] for details).

---

[3] Equation (9) is computationally expensive. r_a is the outermost root of the integrand's denominator and needs to be found with good precision, before the integral can be evaluated.

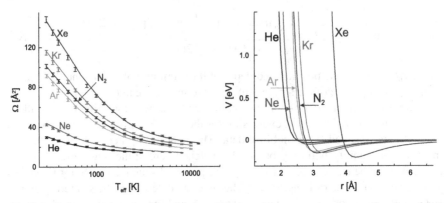

Fig. 7 *Left*: Cross-sections for the scattering of Cs ions from gas molecules as a function of the effective temperature T_{eff}. Data are from [24, 25]. The *solid lines* result from fitting (10) to the experimental data. *Right*: Interaction potentials derived from the cross-sections shown on the *left*

As an example, the procedure of extracting information on the interaction potential will now be illustrated for data on Cs ions. Figure 7 (left) shows experimental cross-sections as a function of the effective temperature for collisions of cesium ions with atoms or molecules of different gases. The data points shown are taken from [24, 25]. The solid lines in Fig. 7 result from fitting the collision cross-sections (10) to the experimental data. The dipole interaction term C_4 was set fixed to the literature values [33] and only the remaining three potential parameters have been adjusted by the fitting procedure.[4] The right panel of Fig. 7 shows the interaction potentials ((7) and (8)) as a function of the distance between the collision partners, employing the potential parameters obtained from the corresponding fits. As to be expected, the onset of the repulsive part of the potential shifts to larger radii for the heavier buffer gases. The position and depth of the potential minimum reflect details in the collision cross-section. Table 1 lists parameters for the interaction potentials of selected ions obtained using the fitting procedure described above [34, 35]. E.W McDaniel and E.A. Mason derived potential parameters from mobility data; these values are supplemented in this survey for comparison.

4.1 Simulation of Ion/Molecule Collisions

With realistic interaction potentials at hand, one can now proceed to microscopically simulate the trajectories of ions in gaseous environments. The idea of simulating the transport of ions in a Monte Carlo fashion was pursued soon after computing power

---

[4] Literature data should be used to reduce the degrees of freedom in the fitting procedure wherever available: Different sets of fit parameters have shown to reproduce the experimental cross-sections equally well, but the ones deviating from the literature C_4, C_6 data tend not to reproduce the K_0-values in mobility simulations.

Table 1 Compilation of parameters for $(n, 6, 4)$ ion-atom potentials for a number of ion/molecule combinations

Ion/molecule	n	r_m (Å)	ε (meV)	γ_q	Reference
Li$^+$/He	12	2.22	47.4	0.10	[33]
Na$^+$/He	12	2.35	40.3	0.15	[33]
Cs$^+$/He	12	2.83	28.0	0.45	[34]
Cs$^+$/He	12	3.36	14.0	0.42	[33]
Hg$^+$/He	14.3	2.48	37.3	0.23	[34]
Cs$^+$/Ne	12.7	2.87	48.4	0.40	[34]
Hg$^+$/Ne	13.7	2.96	42.2	0.37	[34]
Li$^+$/Ar	8	2.35	274	0.30	[35]
Na$^+$/Ar	8	2.69	182	0.84	[35]
K$^+$/Ar	10	2.98	125	0.80	[35]
K$^+$/Ar	12	3.00	121	0.20	[33]
Rb$^+$/Ar	10	3.45	85.1	0.61	[35]
Cs$^+$/Ar	13.2	3.06	130	0.27	[34]
Hg$^+$/Ar	12.6	2.77	242	0.43	[34]
Cs$^+$/N$_2$	12.1	3.12	148	0.40	[34]
Rb$^+$/Kr	12	3.34	119	0.20	[33]
Cs$^+$/Kr	12.2	3.31	146	0.32	[34]
Cs$^+$/Xe	12.8	3.41	200	0.25	[34]
Cs$^+$/Xe	12	3.88	106	0.20	[33]

n, r_m, ε and γ_q are the parameters that define the ion-atom potential as described by (8)

allowed to do the tracking of the ions on a collision-by-collision basis. One of the earliest applications of the technique was the simulation of diffusion coefficients and mobility data [36]. A number of codes have been developed since then to simulate the behavior of ions in dilute gases building on this simulation technique.

The performance of practically all buffer-gas-based RF coolers has been evaluated by microscopic simulations (see, e.g., [34, 37, 38]). These codes use similar algorithms to track an ion in a buffer gas atmosphere; the procedure implemented in the "IonCool" code [34] will be sketched in the following. The code repeatedly processes the following calculation steps, which cover one collision event and the path of the ion to the next collision:

1. Generate the velocity vector of the gas particle before the collision. The vector points in an arbitrary direction, its length follows Maxwell's distribution.
2. The maximum impact parameter b_{max} is calculated. In order to obtain a finite time between two collisions, the range of the collision potential must be limited. One may either directly specify b_{max} or alternatively require a minimum scattering angle θ_{min} to be generated before a collision takes place. For the latter condition, one needs to invert (9) with the available center-of-mass energy in the collision to determine b_{max} from θ_{min}. Specifying a cut-off angle of $\theta_{min} = 10$ mrad or directly limiting b_{max} to 20 Å has yielded good results within reasonable computing time.
3. The impact parameter b is generated following a probability distribution, which increases proportionally to b itself. As motivated above, b is capped by b_{max}.

4. The velocity vector of the ion after the collision is determined. Using (9) the scattering angle in the center-of-mass system is calculated. At this point the code defines the scattering plane, and a transformation back into the lab system yields the velocity vector of the ion after the collision.
5. Calculate the trajectory for the time until the next collision takes place. During a time interval Δt the probability P for a collision increases as

$$P(\Delta t) = 1 - \exp(-v_{\mathrm{eff}} \Delta t / \lambda) . \tag{13}$$

λ designates the mean free path, which depends on the density of the gas particles and the maximum impact parameter. The "effective" velocity

$$v_{\mathrm{eff}} = \sqrt{v_{\mathrm{ion}}^2 + \langle v_{\mathrm{buff}}^2 \rangle} \tag{14}$$

accounts for the contributions of both the ion's instantaneous velocity v_{ion} and the quadratically averaged velocity $< v_{\mathrm{buff}}^2 >$ of the buffer gas atom. If the velocity of the ion changes considerably during the time Δt, step 2 has to be repeated in addition. Values of Δt typically amount to $1\,\mu$s or less.

4.2 Simulation of Mobility Data

One of the simplest applications of the procedure outlined above is the attempt to reproduce experimental mobility data. In such a simulation the ion is allowed to drift through the gas as shown in Fig. 2 (right) long enough so the average drift velocity and thus the mobility can be obtained with sufficiently small variance from the distance traveled in the direction of the constant electric field (cf. (2)). Figure 8 (left) compares simulation results [34] (full circles) and literature mobility data [24] (open symbols) for Cs ions in He, Ar and Xe gas, plotted as a function of the drift velocity v_{d}. The experimental error is quoted as 2–3%. In the simulation the ions had to drift through the gas at room temperature and a pressure of 10^{-3} mbar until 300,000 collisions had happened. This typically resulted in a drift time of several seconds and made the statistical error of the simulation similar to the experimental one. The simulation reproduces the experimental behavior well within the mutual uncertainties.

Considering the computational effort needed to obtain the scattering angles for the $(n, 6, 4)$ potentials, one might wonder how a simpler and easier-to-calculate potential would do. For this reason the simulation was repeated treating the collision partners as hard spheres. The maximum impact parameter in this model is simply the sum of the radii r_{sum} of the spheres representing the ion and the gas molecule. The scattering angle θ can easily be calculated from the impact parameter b: $\theta = 2 \arccos(b/r_{\mathrm{sum}})$. The ionic and atomic radii have been extracted from the online database [39]. Figure 8 (right) displays the results for the hard-sphere model in the same way as for the realistic model (left). In the case of the hard-sphere model

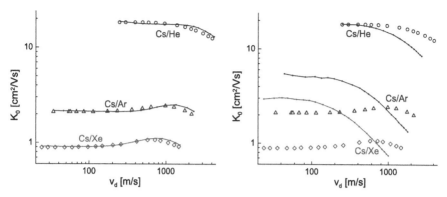

Fig. 8 Comparison of mobility data K_0 from literature (*open symbols*, [24]) with data obtained from simulations (*lines with little filled dots*) for Cs ions in He, Ar and Xe gas as a function of drift velocity v_d. The simulations shown in the *left panel* use a realistic scattering potential while the simulated points on the *right* are based on a hard-sphere collision model

the dependence of K_0 on the drift velocity is not reproduced correctly and the experimental and simulated data sets differ by up to a factor 3 for all used buffer gases. In summary, if one is interested in a fast and possibly more qualitative evaluation of a problem, this model will suffice. However, for quantitative results and comparisons with experimental results the realistic model is recommended.

4.3 Equilibrium Distributions of Buffer-Gas-Cooled Ions – Paul Trap

The cooling process of trapped ions in a gaseous environment will now be illustrated for Cs ions in helium, argon and xenon gas in a 3D Paul trap. The interaction potentials used for the simulation of the individual scattering processes have been obtained as explained before. Figure 9 shows snapshots of the temporal evolution of the points in phase space for one of the radial motions, taken when the confining RF voltage is at its maximum negative value. For the simulation a gas pressure of $p = 5 \cdot 10^{-4}$ mbar and room temperature have been assumed. The operating parameters of the Paul trap are $q_z = 0.03$, $a = 0$, the driving frequency $\Omega/2\pi$ is 1 MHz. At time $t = 0$ (top row) the Cs ions start with an arbitrarily chosen total energy of $10\,\text{eV}$ and randomly distributed phase of the macro motion. The scale of the phase-space figures has been chosen so the initial distribution lies on a circle.

After a certain time equilibrium, distributions are obtained for all three cases (bottom row), which do not change any more if the simulation of the scattering processes is continued. The shape of these distributions depends on the gas temperature and the operating point of the Paul trap. The choice of the collision partners influences both the shape of these distributions and the time it takes to reach them: Whereas it takes 80 ms for Cs^+ in He to reach equilibrium, this already happens

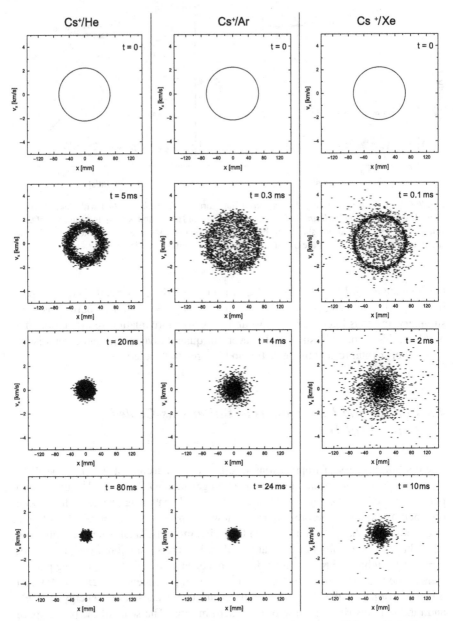

Fig. 9 Simulation of the cooling of a Cs ion ensemble in a Paul trap by collisions with buffer gas atoms (*left*: helium, *middle*: argon, *right*: xenon). The diagrams show the time evolution of the points in phase space for one of the radial motions

at 24 ms for the collision partners Cs^+/Ar and at 10 ms for Cs^+/Xe. Furthermore, it can be seen that the ions' phase-space points for the gases heavier than helium are scattered in a larger area. In a real trap the maximum amplitudes of motion are limited by the trap electrodes, so significant diffusion will cause rapid loss of ions. For this reason heavy buffer gas is not well suited for achieving long storage times, despite the attractively short cooling times it provides.

For the simulation of the ejection of cooled ions from ion traps the knowledge of the shape of the equilibrium distributions is very useful, as this allows one to skip the time-consuming simulation of the cooling process itself. In the case that the mass of the buffer gas molecule is small compared to that of the ion, the equilibrium distributions do not differ significantly for like ions (cf. Fig. 9 bottom: distributions for Cs^+/He and Cs^+/Ar). For these cases the equilibrium density distributions n of the ions in phase space can be predicted easily from a fundamental relationship in statistical mechanics

$$n(x,v) \propto e^{-E(x,v)/(kT)} , \qquad (15)$$

once the total energy $E(x,v)$ is known as a function of the coordinates x and velocities v.

In a Paul trap the total energy of an ion in the macro motion of one dimension $s = x,y,z$ is (per definition) just the same as the energy in an equivalent simple harmonic oscillator: $E_s = 1/2 \cdot m \omega_s^2 (s^2 + (v_s/\omega_s)^2)$. The eigenfrequency ω_s of the macro motion, for not too large Mathieu parameters q_s, is $\omega \approx q_s/\sqrt{8}\Omega$. Noting that (15) separates into similar ones for the three dimensions x,y,z it is easy to see that the equilibrium distributions for any of the three dimensions are of Gaussian shape with standard deviations $\sigma_s = 1/\omega_s \sqrt{kT/m}$ and $\sigma_v = \sqrt{kT/m}$. As already mentioned earlier and illustrated in Fig. 4, the effect of the RF drive can be summarized as a periodic transformation of an initial phase-space distribution into another distribution with the same area. The coordinates and velocities of the macro motion $(s,\dot{s})_{mac}$ transform into the phase-space points of the full motion $(s,\dot{s})$ by a combined stretch-and-rotate action involving a stretch parameter γ_s and a rotation angle θ_r:

$$\begin{pmatrix} s \\ \dot{s} \end{pmatrix} = \begin{pmatrix} \gamma_s & 0 \\ \gamma_s \frac{\Omega}{2}\tan\theta_r & 1/\gamma_s \end{pmatrix} \begin{pmatrix} s \\ \dot{s} \end{pmatrix}_{mac} . \qquad (16)$$

The parameters γ_s and θ_r are complicated functions of the RF phase and the operating parameters and cannot be given in closed form. For practical purposes, however, approximative formulas can be used to determine γ_s and $\tan(\theta_r)$. Assuming an RF voltage with a time dependence proportional to $-\cos(\Omega t)$ and defining the phase of the RF motion as $\xi = \Omega t/2$, useful parametrizations are

$$\gamma_s = 1 + 1.57 \cdot 10^{-3} \cdot e^{8|q|} - 0.531 \cdot q \cdot \cos(2\xi) , \qquad (17)$$

$$\tan(\theta_r) = q \cdot (1 + 0.611|q|^{2.16}) \cdot \sin(2\xi + q \cdot 0.3 \cdot \sin(2\xi)) . \qquad (18)$$

They cover the range $|q| < 0.4$, $0 \le \xi < \pi$, and reproduce the exact values with a deviation of not more than 0.01. The latter two relations, together with (16), allow one to generate the equilibrium distributions for RF-driven Paul-trap-type motions.

For the *linear* Paul trap, the above formulas can be directly applied for the RF-driven transverse motions. The axial motion of ions cooled in the linear Paul trap is essentially a simple harmonic motion in the axial electrostatic potential well. Consequently the equilibrium distribution for the axial motion are time-independent Gaussian distributions with the relevant eigenfrequencies determined by the shape of the axial electric field at the trap center.

4.4 Equilibrium Distributions of Buffer-Gas-Cooled Ions – Penning Trap

The application of buffer gas to cool the motions in a Penning trap is complicated by the fact that the energy of the magnetron motion decreases with the magnetron radius and consequently a straightforward attempt to lower the energy of this particular motion would result in an increase in radius and eventually loss of ion. However, by applying a proper RF quadrupole electric field, the magnetron and the cyclotron motions can be coupled and simultaneously cooled to thermal equilibrium; see, e.g., [40, 41]. As the frequency of the RF drive f_{RF} needs to equal the ion's cyclotron frequency f_c for the coupling to work, this cooling technique is mass selective.

Critical input parameters for the design of a high-efficiency Penning trap mass separator with high resolving power are the widths of the equilibrium distributions that the ions can be cooled down to. We again make use of (15) but now insert the proper expression for the Penning trap's radial motions. Using the customary velocity vectors $V^{\pm}$ [42] defined with the radial coordinates $\rho = (x, y)$ and their time derivatives $d/dt\, \rho \equiv \dot{\rho} \equiv (\dot{x}, \dot{y})$

$$V^{\pm} = \dot{\rho} - \omega_{\mp}(\rho \times \hat{e}_z) \quad \leftrightarrow \quad V_x^{\pm} = \dot{x} - \omega_{\mp} y, \quad V_y^{\pm} = \dot{y} + \omega_{\mp} x, \quad (19)$$

the Hamiltonian can be written as

$$H = \frac{m}{2} \frac{\omega_+ (V^+)^2 - \omega_- (V^-)^2}{\omega_+ - \omega_-}. \quad (20)$$

ω_+ and ω_- denote the frequencies of the cyclotron and magnetron motions, respectively. The solution of the radial equations of motion in a Penning trap as a function of time t in the presence of the coupling RF field with $f_{RF} = f_c$ is somewhat involved (see, e.g., [28]); here only the results will be given:

$$V_x^{\pm} = R^{\pm} \cos(\omega_{\pm} t + \xi^{\pm}) \cos(\frac{k_0 t}{2}) \mp R^{\mp} \cos(\omega_{\pm} t - \xi^{\mp} + \Phi) \sin(\frac{k_0 t}{2}),$$

$$V_y^{\pm} = -R^{\pm} \sin(\omega_{\pm} t + \xi^{\pm}) \cos(\frac{k_0 t}{2}) \pm R^{\mp} \sin(\omega_{\pm} t - \xi^{\mp} + \Phi) \sin(\frac{k_0 t}{2}). \quad (21)$$

The phase of the RF drive enters as Φ; the beat frequency between the two eigenmotions is k_0. The characteristic radii $\rho_{\pm}$ of the two radial motions are introduced through $R^{\pm} \equiv |V^{\pm}| = \rho_{\pm} \cdot (\omega_+ - \omega_-)$; their phases are specified by $\xi^{\pm}$. Inserting (21) into (20) and averaging over time yields the Hamiltonian as a function of the radii $\rho_{\pm}$:

$$H = \frac{m}{4}(\omega_+ - \omega_-)^2 (\rho_+^2 + \rho_-^2) . \tag{22}$$

From (15) (transformed to the radii $\rho_{\pm}$) we can now conclude that the radii $\rho_{\pm}$ are distributed as

$$\frac{dn}{d\rho_{\pm}} \propto \rho_{\pm} \exp\left(-\frac{m(\omega_+ - \omega_-)^2}{4kT}\rho_{\pm}^2\right) . \tag{23}$$

In practice, this result is most useful, as it allows one to generate not only the proper equilibrium distributions for ρ_+ and ρ_-, but – together with a proper randomization of the relevant phases in (21) – also the distributions for the Cartesian coordinates (x, y) and velocities $(\dot{x}, \dot{y})$.

The individual equilibrium distributions for $x, y, \dot{x}, \dot{y}$ can be found by inserting the dependence of $R_{\pm}$ on these coordinates (see above), repeating the time-averaging process and integrating over the coordinates one is not interested in. The results are the Gaussian distributions[5]

$$\frac{dn}{dx} \propto \exp\left(-\frac{m(\omega_+^2 + \omega_-^2)}{8kT}x^2\right) \quad , \quad \frac{dn}{dv_x} \propto \exp\left(-\frac{m}{4kT}v_x^2\right) . \tag{24}$$

Due to the radial symmetry of the problem, the distributions for y and v_y are identical to the ones for x and v_x. The standard deviations of the distributions can be read off as

$$\sigma_x = \sigma_y = 2\sqrt{\frac{kT}{m(\omega_+^2 + \omega_-^2)}} \quad , \quad \sigma_{v_x} = \sigma_{v_y} = \sqrt{\frac{2kT}{m}} . \tag{25}$$

As an example, for singly charged ions of mass $m = 133\,\mathrm{u}$ at room temperature, $\omega_+/2\pi = 500\,\mathrm{kHz}$ and $\omega_-/2\pi = 1\,\mathrm{kHz}$ one obtains $\sigma_x = 87\,\mu\mathrm{m}$ and $\sigma_{v_x} = 194\,\mathrm{m/s}$. This result illustrates that – in the absence of space charge – ion clouds can be cooled down to occupy a volume of much less than a cubic millimeter and have little velocity spread.

5 Applications

Using the techniques discussed in the previous sections many important aspects of the buffer gas cooling of ions in traps can be and have already been addressed. Typical questions include the following:

- How large is the acceptance of the device, i.e., the available phase space for the capture of externally produced ions? This quantity can be obtained by simulating

---

[5] Note that in the expression for dn/dv_x a factor has been neglected, which for all practical cases $(\omega_+ \gg \omega_-)$ is very close to unity.

the injection and cooling process with a representative number of ions that probe the initial phase space.

- What is the capture efficiency of the ion cooler? Comparing the device's acceptance to the phase space of the incoming ion beam allows one to deduce this number.
- How long does it take to cool the ions? As discussed earlier, estimates on the cooling time can be obtained from mobility data for a given electric field strength. However, the presence of varying electric DC and/or RF fields usually requires simulations to obtain more precise answers.
- What is the transverse emittance of the ejected ion beam? The width of the equilibrium distributions allows one to estimate the area of the ion beam in phase space. However, details of the occupied phase space, such as its shape and how it is affected by the extraction optics, can be obtained from Monte Carlo-type ion trajectory calculations starting from the equilibrium distributions.
- What is the axial emittance of the ejected ion beam? For a continuous beam this question reduces to how large the energy spread is. For an ejected ion pulse the shape of the time-of-flight (TOF) distribution is of equal importance. Both energy and TOF distributions are largely determined by the ejection scenario (space and time dependence of the applied fields) and can be affected by the presence of buffer gas in the ejection path (reheating of ions).

All of these questions have been addressed to maximize the performance of the ion cooler and buncher in the low energy and ion trap (LEBIT) experiment [43, 44]. LEBIT converts high-energy fragmentation beams at the NSCL/MSU into low-energy low-emittance beams for high-precision trap-type experiments. A high-pressure gas stopping cell and a radiofrequency quadrupole ion accumulator and buncher manipulate the beam accordingly. The first experimental program to successfully profit from the low-energy beams so obtained is high-accuracy mass measurements of short-lived isotopes with a 9.4 T Penning trap system [45, 46]. The simulation package IonCool [34] was employed to address many aspects in the design and commissioning phase of this facility: ion extraction from the gas stopping cell and transfer into an RFQ mass filter; design and performance of an RFQ ion cooler and buncher; injection of ions into a Penning trap. A typical application of this code, which uses equilibrium distributions of ions and processes collisions of ions with buffer gas atoms on a microscopic basis, is the ejection of ion pulses from the ion buncher. This application will be discussed after a brief overview of the device.

LEBIT's ion accumulator and buncher differs from the standard concept of RFQ ion bunchers in several aspects (details can be found in [4]). The trap part is split into two large sections (cooler and buncher) which operate at different buffer gas pressure. The cooler section is operated at typically $2 \cdot 10^{-2}$ mbar helium to provide fast cooling of ions, whereas the pressure inside the buncher section is kept low at typically $8 \cdot 10^{-4}$ mbar to avoid reheating of ions during ejection. The two sections are linked by the "micro-RFQ," a down-scaled version of the cooler section, which provides differential pumping. Figure 10 shows a cross-sectional view of the buncher section of LEBIT's cooler/buncher. Ions enter the trap section through the

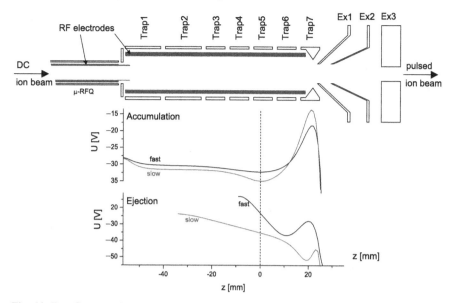

Fig. 10 *Top*: Cross-sectional view of the buncher section of LEBIT's cooler/buncher. The two diagrams *below* show the potentials on the axis of the device present in the accumulation/cooling and the ejection phases for two ejection scenarios. The two ejection settings mainly differ by the electric field strengths applied near the trap center; see text. The position of the trap is at $z = 0\,\text{mm}$

end of the "micro-RFQ" (far left) and then cool and accumulate in the potential well formed by the seven trap electrodes. The two diagrams below show the potentials on the axis of the device present during the accumulation/cooling and the ejection phase. Two ejection scenarios are shown, one for slow and one for fast ion extraction. The main difference is the electric field applied near the trap center, 16.3 V/cm for fast and 3.9 V/cm for slow extraction.

Figure 11a shows two simulated axial phase-space distributions of Ar ions accumulated in the LEBIT buncher in the presence of He buffer gas, extracted as a pulse, accelerated to 2 keV energy and transported to a multi-channel-plate (MCP) ion detector 2.5 m downstream from the buncher. The two distributions correspond to the "slow" and "fast" accumulation and ejection voltage settings introduced above. Equilibrium distributions (as discussed in the preceding section) for room temperature have been used to generate the initial ion distribution for the ejection calculation. The scatter of points around the core of the distributions is an effect of collisions of the ions with buffer gas atoms in the ejection process. Figure 11 also shows time-of-flight (b) and energy (c) profiles projected from the phase-space distribution. At the position of the detector the "fast" ejection scenario produces a very narrow time-of-flight (TOF) distribution at the expense of a broader energy profile; the "slow" ejection produces very little energy spread at the cost of a broader TOF distribution. Figure 11d compares the simulated TOF distribution to experimental data obtained with the MCP detector. The simulation results match the experimental

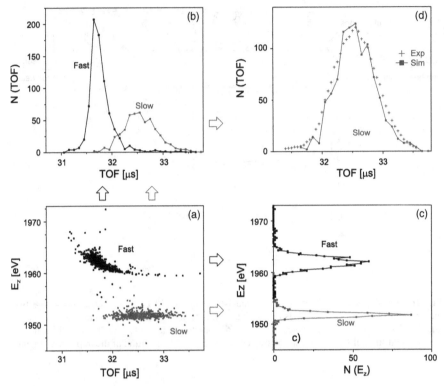

Fig. 11 (a) Simulated axial phase-space distributions of ions accumulated in the LEBIT buncher and extracted as a pulse to the position of an MCP ion detector. The two distributions ("fast" and "slow") result from two different electric field strengths applied in the ejection process. (**b, c**) Projected time-of-flight and energy distributions obtained from the axial phase-space figure. (**d**) The "slow" time-of-flight distribution is compared to experimental results

data very well. All parameters in the simulation (DC and RF voltages, pressure in the buncher section, drift length to the detector) have been set to the experimental values; the only adjustment made is a shift of the measured TOF distribution by 300 ns to account for electronic delays.

References

1. F. Herfurth et al., Nucl. Instrum. Meth. A **469**, 254 (2001).
2. A. Nieminen et al., Nucl. Instrum. Meth. A **469**, 244 (2001).
3. G. Savard et al., Nucl. Instrum. Meth. B **204**, 582 (2003).
4. S. Schwarz et al., Nucl. Instrum. Meth. B **204**, 474 (2003).
5. F. Herfurth et al., Eur. Phys. J. A **25**, 17 (2005).
6. U. Hager et al., Phys. Rev. Lett. **96**, 042504 (2006).
7. G. Savard et al., Phys. Rev. Lett. **95**, 102501 (2005).
8. J. Billowes, Eur. Phys. J. A **25**, 187 (2005)

9. I. Podadera et al., Eur. Phys. J. A **25**, 743 (2005).
10. C. Bachelet et al., Hyperfine Interact. **173**, 195 (2006).
11. R.B. Moore and G. Rouleau, J. Mod. Opt. **39**, 361 (1992).
12. W. Paul, O. Osberghaus, and E. Fischer, Forsch. Wirtsch. Verkehrsministeriums Nordrhein-Westfalen **415**, 1 (1958).
13. S. Schwarz et al., Nucl. Phys. A **693**, 533 (2001).
14. H. Raimbault-Hartmann et al., Nucl. Instrum. Meth. B **126**, 378 (1997).
15. J. Clark et al., Nucl. Instrum. Meth. B **204**, 487 (2003).
16. M. Block et al., Eur. Phys. J. A **25**, 49 (2005).
17. V.S. Kolhinen et al., Nucl. Instrum. Meth. B **294**, 502 (2003).
18. M. Beck et al., Nucl. Instrum. Meth. B **204**, 521 (2003).
19. F. Ames et al., Nucl. Instrum. Meth. A **538**, 17 (2005).
20. D. Habs et al., Hyperfine Interact. **129**, 43 (2000).
21. Poisson Superfish. Source: Los Alamos Accelerator Code Group, www.laacg1.lanl.gov/laacg/
22. D.A. Dahl, Int. J. Mass Spectrom. **200**, 3 (2000). Source: Scientific Instrument Services, Inc., Ringoes, NJ, www.simion.com
23. H.W. Ellis et al., At. Data Nucl. Data Tables **17**, 177 (1976).
24. H.W. Ellis et al., At. Data Nucl. Data Tables **22**, 179 (1978).
25. H.W. Ellis et al., At. Data Nucl. Data Tables **31**, 113 (1984).
26. L.A. Viehland and E.A. Mason, At. Data Nucl. Data Tables **60**, 37 (1995).
27. T. Hasegawa and K. Uehara, Appl. Phys. **B61**, 159 (1995).
28. M. König et al., Int. J. Mass Spectrom. Ion Process. **142**, 95 (1995).
29. F.G. Major and H.G. Dehmelt, Phys. Rev. **170**, 91 (1968).
30. S. Schwarz, Manipulation radioaktiver Ionenstrahlen mit Hilfe einer Paulfalle und direkte Massenmessungen an neutronenarmen Quecksilberisotopen mit dem ISOLTRAP-Experiment, Ph.D. Thesis, Universität Mainz (1998).
31. H.G. Dehmelt, Adv. At. Mol. Phys. **3**, 53 (1967).
32. E.A. Mason and E.W. McDaniel, Transport Properties of Ions in Gases, Wiley, New York (1988).
33. E.W. McDaniel and E.A. Mason, The Mobility and Diffusion of Ions in Gases, Wiley, New York (1973).
34. S. Schwarz, Nucl. Instrum. Meth. A **566**, 233 (2006).
35. P. Schmidt, REXTRAP – ion accumulation, cooling and bunching for REX-ISOLDE, Ph.D. Thesis, Universität Mainz (2001).
36. H.R. Skullerud, J. Phys. B **6**, 728 (1973).
37. T. Kim, Buffer gas cooling of ions in an RF ion guide: a study of the cooling process and cooled beam properties, Ph.D. Thesis, McGill University, Montreal (1997).
38. G. Ban et al., Nucl. Instrum. Meth. A **518**, 712 (2004).
39. M. Winter, Online database, data available at www.webelements.com
40. G. Bollen et al., J. Appl. Phys. **68**, 4355 (1990).
41. G. Savard et al., Phys. Lett. A **158**, 247 (1991).
42. L.S. Brown and G. Gabrielse, Rev. Mod. Phys. **58**, 233 (1986).
43. S. Schwarz et al., Nucl. Instrum. Meth. B **204**, 507 (2003).
44. P. Schury et al., Eur. Phys. J. A **25**, 51 (2005).
45. R. Ringle et al., Eur. Phys. J. A **25**, 59 (2005).
46. G. Bollen et al., Phys. Rev. Lett. **96**, 152501 (2006).

Highly Charged Ions and High-Resolution Mass Spectrometry in a Penning Trap

Sz. Nagy, K. Blaum, and R. Schuch

1 Preface

Mass spectrometry is the *art* of measuring the charge-to-mass ratio, q/m, of atomic or molecular ions. Any device capable of producing a mass spectrum can be called a *mass spectrometer*. Mass spectrometry has more than 100 years of history, which is rich in celebrated discoveries, Nobel Prize laureates, and great technical inventions [1, 2]. Among the paramount discoveries facilitated by mass spectrometers are the discovery of the electron [3, 4], the discovery of isotopes [5], and the discovery of the so-called mass defect [6, 7]. Mass spectrometry while having its roots in physics branched into chemistry and in the recent decades became one of the most important tool in biochemical sciences [8].

The history of atomic mass spectrometry began with J.J. Thomson (1856–1940) at the Cavendish Laboratory of the University of Cambridge. In 1897, Thomson discovered the electron and measured its charge-to-mass ratio [3]. The apparatus built by Thomson employed aligned magnetic and electric fields having ions of the same species focused on a photographic plate as parabolas. The resolving power of Thomson's first *parabola spectrograph* was around $R = M/\Delta M = 10$–20 [9]. The pioneering work of Thomson was followed by a long series of improvements; a number of different magnetic mass spectrometers combined with electrostatic devices were developed [10, 11, 12, 13, 14], and the resolving power of the mass spectrographs/spectrometers was increased in average by one order of magnitude per decade, see Fig. 1.

A breakthrough came at the end of the 1950s when L.G. Smith [15] built his radio-frequency *mass synchrometer* [16], which was the first mass spectrometer to

Sz. Nagy and R. Schuch
Atomic Physics, AlbaNova, Stockholm University, S-106 91 Stockholm, Sweden

Sz. Nagy and K. Blaum
Department of Physics, Johannes Gutenberg-University, Staudingerweg 7, D-55128 Mainz, Germany

Nagy, Sz. et al.: *Highly Charged Ions and High-Resolution Mass Spectrometry in a Penning Trap.*
Lect. Notes Phys. **749**, 119–154 (2008)
DOI 10.1007/978-3-540-77817-2_5

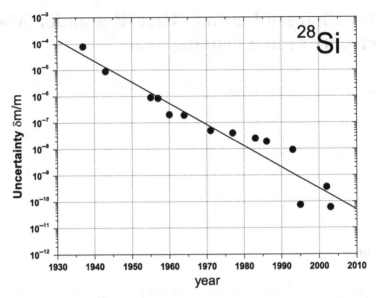

Fig. 1 The precision of the ^{28}Si mass was improved in average about one order of magnitude per decade [2], which illustrates the tremendous progress in the field of high-precision mass spectrometry. The last three points are measured by Penning-trap mass spectrometers

apply a frequency measurement. The resolving power achieved by Smith [17] could be significantly improved only by introducing an entirely new technology. This occurred after Penning traps were introduced [18] making it possible that in 1980 the first Penning-trap mass measurement using the time-of-flight ion cyclotron resonance technique could be performed [19]. Today, the most accurate mass measurements are accomplished by comparing cyclotron frequencies in a Penning trap using single ions at the WS-PTMS, Seattle [20], single ions and molecules at FSU[1] [21, 22], Tallahassee, highly charged ions at SMILETRAP [23], MSL, Stockholm, or radioactive ions at ISOLTRAP [24], ISOLDE/CERN, Geneva, and several other facilities, see [25] for a broad overview.

The SMILETRAP facility is unique due to the usage of highly charged ions (HCI). In fact, it is the only experiment where the benefit of using ions in high charge states is exploited for reaching high precision in mass measurements. Thus, SMILETRAP will be presented as a model experiment for high-precision mass measurements with HCI. But, first let us get more familiar with highly charged ions and their creation/production techniques.

---

[1] In May 2003, the MIT ICR Lab was closed and the apparatus was moved to Florida State University (FSU), Tallahassee, where it has been set up in a new ICR laboratory by E.G. Myers et al.

2 Brief Introduction to Highly Charged Ions

Ionized states of matter are of interest for questions in fundamental physics as well as for applications in other fields of physics and technology. More than 90% of matter in the universe appears in an ionized state. Almost all information in astronomy and astrophysics comes from radiation of electrons and ions, reacting in the plasma of cosmic objects, e.g., in the Sun and other hot stars where the charge state can be as high as $q = 30$ [26]. Photon emission spectra from HCI are observed from supernovae and even in comets passing close to the earth. Studies of the optical spectra give information of the elemental composition and the temperature of such objects. Highly charged ions are, however, surpassingly rare on Earth. They can be found in the upper atmosphere as a result of the ionization of atoms and molecules by cosmic rays or solar wind. Since the Earth's atmosphere is mostly composed of light atoms, nitrogen and oxygen, the charge state of these ions is low, only about $q = 1$–5. Similar charge states are also present in some electric discharge plasmas, e.g., thunderstorm lightning. In tokamak-plasma, highly charged ions occur as impurities and therefore it is important to know their properties. Accurate knowledge of the properties of HCI is also necessary for the diagnostics of these hot plasmas for fusion energy research, since theoretical methods currently do not reach the precision achieved in experiments [27, 28].

HCI became the subject of extensive scientific studies especially after finding efficient ways to create them in a laboratory environment. The specific properties distinguishing a highly charged ion from a neutral atom are mainly due to the unshielded Coulomb field of the nucleus of the ion. Its effects can be seen in the structure of the ion and in the interactions involving other atoms or electromagnetic fields. HCI provide a simple system in which electrons are subject to the strongest electric fields available. The electric field experienced by a lonely electron orbiting a uranium nucleus is approximately $10^{16}\,\mathrm{V\,cm^{-1}}$, stronger than any static electric field available in any laboratory. Even the most intense laser fields are several orders of magnitudes weaker than this, as shown in Fig. 2. The conditions inside HCI thus provide a unique testing ground for quantum electrodynamics (QED) in strong fields [29]. This allows studies of atomic properties in the ultra-relativistic regime, thereby exploring the validity range of QED.

HCI can also give tremendous benefits in precision mass measurements. In mass spectrometry with Penning traps, the precision increases linearly with the charge of the ion, a feature that was exploited in the Penning-trap mass spectrometer SMILETRAP. Unfortunately, one pays a price for this precision increase: in order to determine the atomic masses, one has to correct the measured ionic mass for the neutralization energies of the ionized electrons. In the case of HCI, two types of potential energy are distinguished: the ionization energy and the neutralization energy. For a singly charged ion, these two are identical in magnitude, but for HCI they can differ by more than a factor of 10. Ionization energy is the energy required to remove one electron, producing the charge $q + 1$ from an ion with charge q. Neutralization energy is the energy released when adding back all of the missing electrons. The

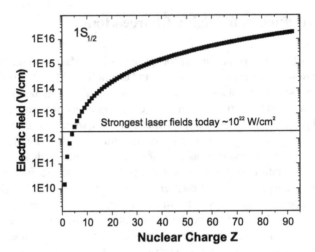

Fig. 2 The approximate value of the electrical field strength experienced by an electron bound in the $1S_{1/2}$ state of different hydrogen-like ions in the range $Z = 1$ to $Z = 92$. Calculated by simple Z^3 scaling

ionization energy is most relevant for the production of the ions, which typically occurs in a stepwise process (one electron is removed after another).

Let us consider a single atom of a certain chemical element. The nucleus, where most of the mass is concentrated, is composed of N neutrons with mass m_n and Z protons with mass m_p. The nucleus is surrounded by Z electrons. The mass of the atom is somewhat less than the sum of the masses of its ingredient particles; the difference is called the mass defect [5], and it is explained by the binding energies of the nucleons and electrons:

$$M_{atom} = N_n \cdot m_n + Z \cdot m_p + Z \cdot m_e - E_b^{nuclear} - E_b^{atomic} , \qquad (1)$$

where Z is called the nuclear charge number, $E_b^{nuclear}$ is the total nuclear binding energy, and E_b^{atomic} is the total atomic binding energy.

3 Production of Highly Charged Ions

Highly charged ions can be produced in various ways, which may involve huge accelerator facilities [30], table-top devices [31], or even miniature devices [32, 33]. The physical process leading to ionization is usually collision of atoms with charged particles (usually electrons) or interaction with electromagnetic radiation. Moderately charged ions with appreciable intensities can be obtained, e.g., from a penning ionization gauge (PIG) [34] source, an electron cyclotron resonance (ECR) [35] ion source, or a laser-produced ion source (LPIS) [34]. For the PIG, ECR, and LPIS, the attainable charge states are approximately limited at 20+. These ion sources are also used for high-precision mass measurements with Penning traps. But for creating the highest charge states of an ion, two methods are most useful. A multi-stage accelerator facility can be run in an accel–decel mode [36, 37, 38]. In this case, a high-energy ion beam is sent through a gas or thin solid foil target and the stripped HCI are injected into a decelerator, which can be a HF accelerating structure or a synchrotron storage ring. One should fulfill the Bohr criterium for getting efficient stripping ($v_{ion} \approx v_n$, where v_{ion} is the ion velocity and v_n is the average Bohr velocity in the orbital with quantum number n) [36]. One needs also to apply a suitable thickness of the stripping target which is determined by the ionization cross-section of the given shell at the given beam energy. For example, within the HITRAP [39] project at GSI, one aims at stripping ions up to uranium completely (this requires an energy of around 500 MeV/u), then deceleration in several steps, and finally trapping the HCI basically at rest. The future physics program at this facility includes high-precision mass measurements, for further details see [39].

A rather compact and relatively inexpensive device also capable of fully ionizing all atomic species up to U^{92+} is the electron beam ion source (EBIS) [40] and its modified shorter version the electron beam ion trap (EBIT) [41, 42, 43]. From the EBIT also, HCI can be extracted, so it can operate as a source [42]; we denote, therefore, in the following both types of sources by EBIT/S. The EBIT/S are capable of producing pulses of HCI of low emittance and low velocity, offering also good control and tunability. They have proven to be very powerful machines for high-precision Penning-trap mass spectrometry experiments.

3.1 The Electron Beam Ion Trap and Source

The EBIS principle was proposed by Donets in 1969 [40], and ever since it has been very successful in producing highly charged ions for a broad range of physics experiments [44]. In Stockholm, a cryogenic EBIS named CRYSIS (CRYogenic Stockholm Ion Source) [45, 46] has supplied the SMILETRAP Penning-trap mass spectrometer with HCI, until recently. It is out of operation now and a new SMILE-TRAP is being set up at an EBIT at AlbaNova in Stockholm. We will describe both the EBIS and EBIT features for production of HCI for a Penning-trap mass spectrometer.

Nowadays, an electron beam ion source called REXEBIS is being used for charge breeding [47] at the online isotope separator facility ISOLDE [48], at the REX-ISOLDE [49] post-accelerator at CERN, see Fig. 3. An EBIT is also being set up for charge breeding at the online isotope separator facility TITAN [51] at TRIUMF.

In the top part of Fig. 4, the principle parts of an EBIS are shown in a schematic way. An electron beam is produced by an electron gun at one end and then it is compressed to high density by a strong solenoidal magnetic field, typically 2–6 T. The electron beam passing through the solenoid is decelerated and stopped in the electron collector. The EBIS trap region is built up by a series of co-linear cylindrical drift tubes over typically a meter length. In CRYSIS, it is a series of 33 drift

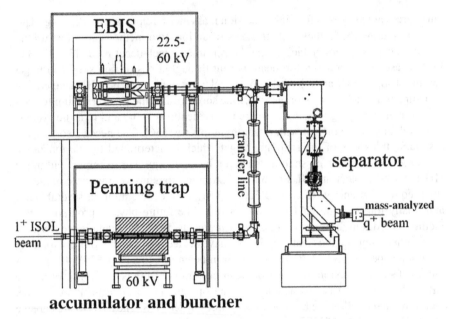

Fig. 3 The layout of the Penning-trap accumulator and EBIS charge breeder of REX ISOLDE at CERN [49]. Radioactive 1+ ions produced at ISOLDE [48] are accumulated and phase-space cooled in the Penning trap and thereafter injected into the REXEBIS for charge breeding [50]

tubes spanning approximately 1.5 m. In an EBIT, there are usually only three short cylindrical drift tubes with only a few centimeters length. An important difference to the EBIS is that the magnetic field in an EBIT is made by coils in Helmholtz geometry. This and slits in the central drift tube allows observation of photons from the trapping region, which is very important for experiments in the trapping mode of EBIT. The cryostat chamber together with the superconducting magnet coils, is cooled to liquid helium temperature (4.2 K). It is configured to provide the extremely good vacuum, typically below 10^{-11} mbar inside the drift tubes, due to the cryogenic environment which is necessary to prevent electron capture of HCI from rest gas. Electrostatic axial barriers for ions are produced at the ends of the trap region by applying positive voltages on the end electrodes. Ions are confined radially by the space charge of the electron beam and the magnetic field.

There are two ways of seeding an EBIT/S. For the production of noble gas ions and ions from molecular gases with a low melting point, the neutral gas is injected into the trap. With an EBIS, this is done simply by injecting the gas into the region of the injection trap (Fig. 4 Injection, the first trapping region on the right). At an EBIT, one makes an atomic/molecular beam by two apertures and a drift region in between with a differential pumping stage. The pressure in the drift region should be below 10^{-5} mbar in order to have a mean free path of several meters, and particles in that beam will cross the electron beam in the central drift tube where the particles

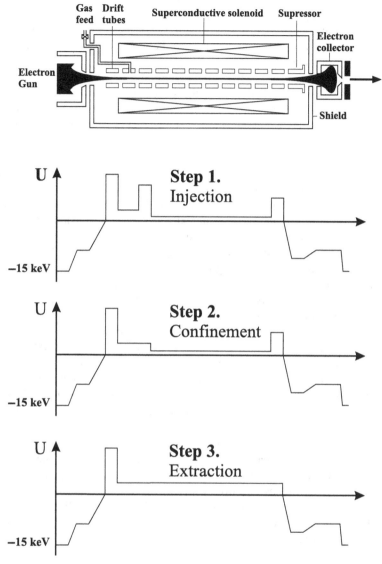

Fig. 4 A sketch of the electron beam ion source CRYSIS [46] and the three major steps of operation. For details see text

get ionized and trapped. Besides the available gases, also liquids or metals can be evaporated and injected.

The second way is far more advanced and it involves the use of an external ion injector. At EBITs, a metal vapour vacuum arc (MEVVA) source is frequently used [52, 53]. It produces large abundances of metal ions in charge states 1+ to 3+. The external ion injector at CRYSIS was the composition of a commercial cold or

hot reflex discharge ion source (CHORDIS), manufactured by DANFYSIK) on a
20 kV platform and an isotope separator with a high-mass resolving power. Here,
a wide variety of ions of different isotopes can be produced through gas injection,
sputtering, or evaporation from an oven.

A critical limitation of ion and gas injection is isotopically pure ion beams requested by spectroscopic experiments and also for precision mass measurements.
As the trap of an EBIT/S can only be filled up to about 20% of the electron beam
space limit, it is not very economical to inject ions that are of no interest. It is thus
necessary to inject either isotopically enriched gas atoms, which can be very expensive, or singly charged ions cleaned by isotope separation.

After capturing the ions in the trap, they are ionized for typically 100 ms to several seconds, where the confinement time and the energy of the electron beam are
chosen according to the requested charge state. At the end, the voltage of one of the
end electrodes is lowered and the ions are extracted. They pass axially through the
hollow electron collector and into the beam transport line. The charge-state evolution can be rapidly monitored using a time-of-flight technique and the desired charge
state then selected by a $90°$ analyzing magnet. An important feature of an EBIT/S
is that it produces a fixed amount of positive charges per pulse. When delivering
beams to the SMILETRAP experiment, a typically 100 μs long ion pulse is extracted
at the end of each production cycle, containing about 10^8 charges, see Table 1. The
extraction is done at 3.4 kV. The energy spread of the beam is rather small, typically 5–10 eV, which is estimated from the ion capture in the trap of SMILETRAP.
A pulse of 100 μs length cannot be entirely captured in a Penning trap. Most of the
ions transferred are lost. Thus, for re-trapping, the pulse length should be optimized
to a shorter length without loosing the energy definition. At the EBIT at AlbaNova,
Stockholm, a pulse length of <100 ns was reached [54] which could be re-trapped
almost completely.

Another essential feature of an EBIT/S is that the well-defined electron beam
energy and ionization time lead to a narrow charge-state distribution. The attainable
charge state is limited by the electron beam energy which must be greater than the
ionization energy for the last electron. CRYSIS was designed for 50 keV electron
energy, which is sufficient for the production of bare nuclei of all elements up to
xenon. The highest energy electron beam of around 200 keV was reached in the
Super-EBIT in Livermore [55, 56].

Table 1 Typical CRYSIS parameters used in the case of mass measurements at SMILETRAP

Parameter	Value
Electron beam energy	14.5 keV
Electron beam current	70–145 mA
Magnetic field	1.5 T
Ion energy	3.4 keV $\times q$
Charge per pulse	0.5–2 nC
Ion pulse length	100 μs
Confinement time	20 ms–2 s
Maximum charge state	$^{204}Hg^{52+}$

During ionization, the ions are heated by the ionizing electrons which results in ion losses, large emittance of the extracted beam, and reduced ionization probability. Ion cooling during confinement is thus very important. In an EBIT/S, this is done by injecting a gas of lighter atoms than those that should be ionized. The lighter ions reach a lower charge state and can thus evaporate from the trap, cooling the remaining heavier ones. Compared to an EBIS, evaporative ion cooling can be much more effectively done in an EBIT. The evaporation of the light ions occurs axially. Due to the strong axial magnetic field, it is impossible to evaporate ions radially. But a 1.5 m long solenoid of an EBIS, such as CRYSIS, makes axial evaporation very ineffective. Only a small fraction of ions that escape through the end caps of the trapping region in an EBIS can contribute to cooling. Still it was observed that evaporative ion cooling has an effect on the ion temperature for ions coming from CRYSIS. With EBIT, light ion cooling is very effective, as the trap is only 2 cm long. This has been established in Livermore and is used now at all these machines throughout the world.

4 Mass Measurement of Highly Charged Ions in a Penning-Trap Mass Spectrometer

The Penning trap, see Fig. 5, can be used as a mass spectrometer due to the fact that the oscillation frequency of the trapped ions in a magnetic field is mass dependent. A comprehensive description of Penning traps can be found in the book of Ghosh [59] and in a recent book of Major et al. [60]. Most aspects of ion motion in a Penning trap have been described by Brown and Gabrielse in [61]. Here only

Fig. 5 Two realizations of a Penning trap, a hyperbolical (*left*) and a cylindrical (*right*) version. In both cases, the electrodes define a quadrupole potential [57, 58]

a brief introduction will be given (for more details, see contribution by G. Werth in this lecture series), necessary for understanding the HCI mass measurement procedure at SMILETRAP. From the motional frequencies of a trapped ion, it is possible to determine the free cyclotron frequency, which is related to the ion mass via the following equation:

$$v_c = \frac{1}{2\pi} \frac{qB}{m} , \qquad (2)$$

where q/m is the charge-to-mass ratio of the stored ion and B is the magnetic field strength. This frequency can be determined either by measuring the image currents in the trap electrodes induced by the oscillating ions [20, 62] or by a time-of-flight technique [19, 63]. At SMILETRAP, the latter one is applied which thus shall be described in more detail.

4.1 Principles of Mass Measurements Using the Time-of-Flight Ion Cyclotron Resonance Technique

The time-of-flight (TOF) technique was proposed by Bloch [64] and first applied to precision mass measurements in 1980 by Gräff et al. [19]. It is a destructive technique in the sense that one loses the trapped ions in each detection cycle.

In the ideal case, the three different oscillations (magnetron oscillation: ω_-, axial oscillation: ω_z, and modified cyclotron oscillation: ω_+) of a highly charged ion inside a Penning trap are uncoupled and can be described by a quantized harmonic oscillator, as shown in [59, 65]. An external electric radio-frequency field (RF) can be used to enhance the energy of each individual motion by resonant excitation at the eigenfrequency. Typically, dipolar and quadrupolar excitations are used [24, 66]; octupolar is also possible [67, 68]. For mass determination purpose, a coupling of the magnetron motion to the reduced cyclotron motion is achieved by a quadrupolar driving field at frequency $\omega_{RF} = \omega_c = \omega_- + \omega_+$, which is applied simultaneously on the two pairs of opposite segments of the ring electrode as illustrated in Fig. 6.

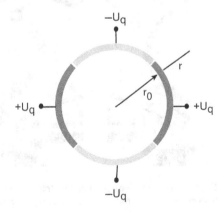

Fig. 6 Transverse cut of a four-fold segmented ring electrode of a Penning trap showing the connections for an azimuthal quadrupolar excitation used in the case of mass measurements. The voltage applied to one pair of opposite segments is phase-shifted by 180° with respect to the voltage applied to the other pair. In this way, an electric quadrupole field is generated in the radial plane

The electric quadrupole field used for excitation can have the form:

$$E_q = U_q \cos(\omega_q t - \phi_q) \cdot \begin{bmatrix} x \\ -y \\ 0 \end{bmatrix} . \tag{3}$$

The azimuthal quadrupole excitation leads to a coupling of the cyclotron and magnetron motions, which means that a continued excitation at the resonance frequency, ω_c, will result in a periodic conversion from an initial pure magnetron motion into a pure cyclotron motion, i.e., beating between the two motions [66] takes place. The beating frequency or Rabi flopping frequency, can be written as [66]

$$\omega_b = \frac{U_{RF}}{4Br_0^2} \left(\frac{\omega_c}{\omega_+ - \omega_-} \right) . \tag{4}$$

It should be noticed that, to the first-order approximation, ω_b is independent of the ion mass since $\omega_c \approx \omega_+ - \omega_-$. The time period T_b for one full oscillation is $T_b = 2\pi/\omega_b$. After an excitation time of $T_b/2$, an initially pure magnetron motion is completely transferred into a pure cyclotron motion; at this time, the modified cyclotron motion has a large radius equal to the radius of the initial magnetron motion. This leads to a gain in the radial energy, which later is converted into axial energy in the fringe field of the magnet.

The cyclotron frequency is obtained by scanning the frequency of the excitation signal and measuring after ejection the time of flight of the ions flying from the trap to a detector located outside the strong magnetic field, as illustrated in Fig. 7. The excitation enhances the radial kinetic energy due to the coupling and conversion of the radial motions, since $\omega_c \gg \omega_-$. The calculated radial energy gain for a 1 s quadrupolar excitation as a function of the frequency detuning, $\omega_{RF} - \omega_c$, is shown in Fig. 8 (left). The radial energy will be converted into axial energy in a decreasing non-linear B-field after the ion is ejected out of the trap, see Fig. 7. The flight time of the ions can be calculated in the following way [66]:

$$T(\omega_{RF}) = \int_{z_0}^{z_1} \left\{ \frac{m}{2[E_0 - q \cdot V(z) - \mu \cdot B(z)]} \right\}^{1/2} dz , \tag{5}$$

where E_0 is the total initial energy of the ion and $V(z)$ and $B(z)$ are the electric and magnetic fields along the ions path from the trap at z_0 to the detector at z_1 (see Fig. 7). In the resonance case ($\omega_{RF} = \omega_c$), the magnetic moment $\mu = (qe\omega_c r_c^2)/2$ (r_c is the radius of the cyclotron motion after excitation) has its maximum value. By scanning the excitation frequency and recording the ion's flight time to the detector, the resonance can be observed as a well-pronounced minimum in the time-of-flight spectrum. The calculated time-of-flight cyclotron resonance spectrum for a 1 s quadrupolar excitation is shown in Fig. 8 (right). It can be shown [69] that the theoretical full-width at half-maximum (FWHM) is

$$\triangle v_c \approx \frac{0.8}{T_{RF}} . \tag{6}$$

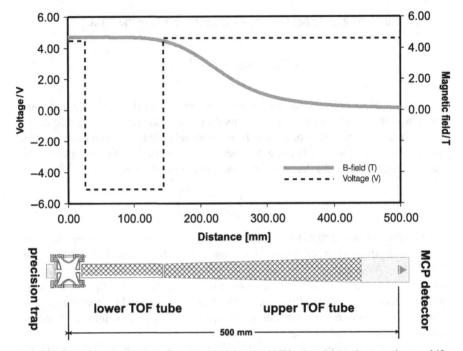

Fig. 7 The B-field variation along the time-of-flight tube (*full line*) and the voltage on the two drift tubes (*dashed line*) are plotted as a function of the distance to the detector

To obtain the center value, which is the mass-dependent cyclotron frequency, ν_c, the measured resonance curve is fitted using a least-squares method with the theoretical lineshape. In this way, the center value can be obtained typically to $\sim 1\%$ of the FWHM.

4.2 Advantages of Highly Charged Ions in Penning-Trap Mass Spectrometry

To understand the benefits of using HCI for high-precision mass spectrometry, it is sufficient to look at (2). A higher charge state q results in a higher cyclotron frequency; therefore, a higher precision can be achieved, since $\delta m/m = \delta \nu/\nu$. The statistical uncertainty is given by

$$\left(\frac{\delta m}{m}\right)_{\text{stat}} \propto \frac{m}{q T_{\text{RF}} B \sqrt{N}} . \tag{7}$$

In Fig. 9 (left), the cyclotron frequency is calculated for all charge states q of $^{238}\text{U}^{q+}$ in a $B = 7\,\text{T}$ magnetic field using (2). The frequency increases linearly with the ionic charge state q from about 500 kHz to about 40 MHz.

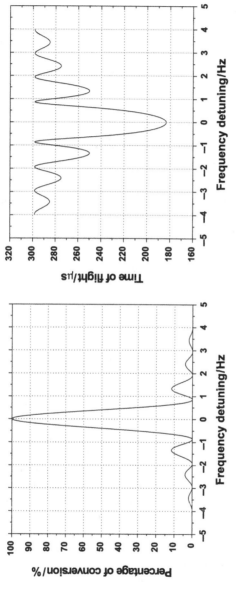

Fig. 8 *Left*: Theoretical calculation for the radial energy gain after a 1 s quadrupolar excitation. *Right*: The corresponding calculated time-of-flight cyclotron resonance spectrum as a function of the frequency detuning

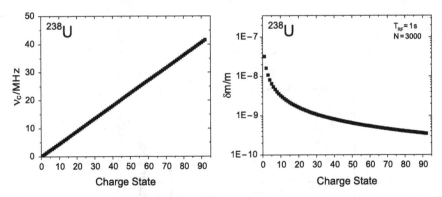

Fig. 9 *Left*: The cyclotron frequency, ν_c, of ^{238}U as a function of the charge state q in a magnetic field $B = 7$ T. *Right*: The calculated statistical uncertainty of an assumed mass measurement as a function of the charge state using $N = 3000$ ions and an observation time of 1 s. Compared to 1+ ions, by using 92+ ions two orders of magnitude can be gained in mass precision

In Fig. 9 (right), the statistical uncertainty is plotted as function of the charge state for $^{238}U^{q+}$, calculated with (7), using $N = 3000$ ions, $B = 7$ T, and an observation time $T_{RF} = 1$ s. One can see that by increasing the charge state to $q = 92$, the statistical uncertainty changes from 3×10^{-8} to 3.5×10^{-10}; thus up to two orders of magnitude can be gained by using highly charged ions in this case.

5 Cooling of Highly Charged Ions

Cooling of HCI in a Penning trap means a reduction of the motional amplitudes of the HCI oscillating in the trap potential. In general, the attainable precision in Penning-trap mass spectrometry depends strongly on the possibility of ion cooling. Cold ions are located in potential minimum (center) of the trap, i.e., are confined in smaller volume, thus probe less of the field imperfections. At the same time, the observation time can be extended; thus, higher precision can be reached. The well-developed buffer-gas cooling technique [70, 71] cannot be applied to HCI due to charge exchange losses. The other cooling mechanism of charged particles in a strong magnetic field by losing their kinetic energies via synchrotron radiation is not efficient. The rate of the synchrotron radiation is inversely proportional to the particles mass cubed, i.e., the rates for HCI are more than 10^{10} times slower than those for electrons and positrons. Therefore, only electrons and positrons can be cooled via synchrotron radiation within some reasonable time with today's strongest magnetic fields available by using state-of-the-art superconducting magnets. Heavy HCI can accordingly be cooled sympathetically if they are stored simultaneously with electrons [72] and/or positrons [73] in a strong magnetic field. In principle, cooling of HCI with positrons is better than with electrons because HCI cannot recombine with positrons and do not change charge during the cooling procedure. On the other hand, the number of electrons (10^9——10^{10}) in a trap can be practically a few orders of magnitude larger than the number of positrons (10^7—10^8). A strong positron source in addition is costly, and the electron cooling scheme has the large advantage that it has a short cooling time relative to the lifetime due to losses by capture from residual gas.

For precision mass measurements, only a single HCI is needed and wanted at a time in the trap. If one starts with a large number, then the ion energy distribution can also be reduced by evaporative cooling. For that purpose, one uses a cooler Penning trap that can store a large number of highly charged ions. By lowering the trapping potential, a certain amount of trapped ions can be evaporated and then the potential is raised again to equilibrate the remaining ions to a lower temperature. This happens effectively within 1 ms by collisions between the ions. Simulations have shown that several of these evaporation cycles should decrease the ion temperature of the remaining ions to below eV energy.

One may think of direct laser cooling, but this is unfortunately not feasible for HCI since there are no electric dipole transitions accessible with laser light. A possible scheme is to merge the HCI with a laser-cooled ion cloud (Be^+, Mg^+) and cool the HCI by Coulomb collisions (sympathetic cooling) to the temperature of the singly charged ions. Cooling of HCI by such a method was demonstrated at the Livermore EBIT [74, 75], and cooling with Mg^+ ions is foreseen for the MLLTRAP in Munich [76]. In-trap preparation of HCI for precision mass measurements by cooling in a strongly coupled plasma of laser-cooled $^{24}Mg^+$ ions is investigated by molecular dynamics simulations [76] and with Be^+ ions in [77].

Electron cooling of p and $\bar{p}$ was invented in Penning traps by G. Gabrielse [78] and is since used at ATRAP [79] and ATHENA [80] at CERN. The electron cooling of HCI has been investigated and evaluated for the HITRAP project [81]. However, it is only applicable at energies of the HCI higher than $100\,eV/q$ because at lower energy the recombination rate becomes too high [81]. The other well-developed method is resistive cooling of ions in a Penning trap [82, 83]. The cooling time of this method decreases when the charge state increases. However, to capture the ions with a high-energy spread requires the use of a large catcher trap, and the cooling time increases with the square of the trap size. Electron cooling and resistive cooling will be briefly described in the next two sections.

5.1 Electron Cooling

Electron cooling is a well-established technique to increase the phase-space density of ion beams in storage rings [84, 85]. This is achieved by aligning the ion beam with a cold dense electron beam from a cathode. Electron cooling in a Penning trap has first been employed at LEAR/CERN to slow down antiprotons from an energy of 3 keV to rest [78]. It is now one of the key processes that enabled the first production of antihydrogen atoms at CERN [79, 86].

The difference between cooling of HCI and antiprotons is that for HCI recombination processes have to be taken into account. A recent theoretical investigation of

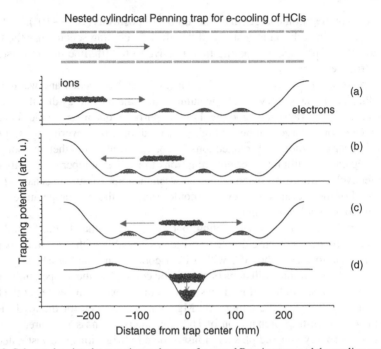

Fig. 10 Scheme showing the experimental setup of a nested Penning trap and the cooling cycle for electron cooling of highly charged ions at HITRAP [39]. The ions enter the trap (**a**), are reflected and caught after their first turn by switching the potential (**b**), cooled by the electron cloud (**c**), and finally by resistive cooling (**d**) after which they will be released from the cooler trap in order to be transferred to the precision experiments

electron cooling of HCI can be found in [72] and also within this volume. In a Penning trap the magnetic field confines the charged particles in the direction perpendicular to the magnetic field lines and the electrostatic field in the direction parallel to the magnetic field lines. By using a cylindrical nested Penning trap, as foreseen in the HITRAP experiment [39], both electrons and ions can be stored in the same trap, see Fig. 10. First, electrons are loaded into a local potential well. A decelerated bunch of about a few hundred ions enters the trap at an energy of about $10\,keV/q$. Entering the nested Penning trap, HCI interact with the electrons by Coulomb interactions. These collisions lead to a frictional force that decelerates the ions. The trapped ions can be slowed down in repeated collisions with the trapped electrons until their energy is about $100\,eV/q$. After electron cooling, the highly charged ions are spatially separated from the trapped electrons in the cylindrical nested trap configuration in order to minimize recombination losses, and resistive cooling sets in.

5.2 Resistive Cooling

With resistive cooling, the motional energy of the stored charged particles is damped by use of an external circuit that is continuously kept in resonance with the

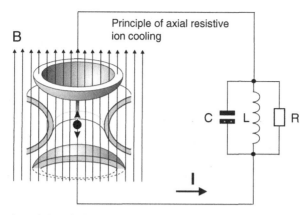

Fig. 11 Illustration of the principle of the resistive cooling technique. The energy of the axial motion of an ion stored in a Penning trap can be dissipated through the resonant impedance

eigenfrequency of the ions' oscillation. The kinetic energy of the ions is dissipated in the impedance of the circuit via the image currents induced in the trap electrodes [82, 83]. Finally, the energy of the ions corresponds to the temperature at which the impedance is kept (generally the temperature of liquid helium, 4.2 K), i.e., a thermal equilibrium with the environment is reached.

Resistive cooling is especially efficient for ions with a large charge-to-mass ratio q/m. A possible experimental configuration for resistive cooling of the axial ion motion with an external resonant circuit is shown in Fig. 11. The advantage of the resistive cooling is that no particles should be lost in the process.

6 The Penning-Trap Mass Spectrometer SMILETRAP

The SMILETRAP (*S*tockholm-*M*ainz-*I*on-*LE*vitation – *TRAP*) was started as a collaboration between the Manne Siegbahn Institute (MSI) at Stockholm University, Sweden, and the Physics Department of the Johannes Gutenberg University in Mainz, Germany. The main objective was to perform mass measurements relevant for fundamental physics exploiting the precision gain by the use of highly charged ions [87]. The construction started in Mainz in the summer of 1990, and in December 1991, the first cyclotron resonance spectra were recorded. The apparatus was moved to Stockholm and connected to the electron beam ion source (CRYSIS) at MSL in 1993 [45, 46]. It was, after many years of successful operation, turned off and dismantled in the beginning of 2007. A new SMILETRAP mass spectrometer is presently being set up at the AlbaNova Physics Center in Stockholm using a fully refrigerated EBIT. A schematic layout of the new SMILETRAP II facility is shown in Fig. 12. Isotopically clean pulses of singly charged ions are injected into the refrigerated EBIT ion source, R-EBIT, and after charge breeding, a highly charged ion pulse is extracted, charge separated in an analyzing magnet. First, the ions are stopped and retarded in a preparation trap and later sent to a precision trap for mass

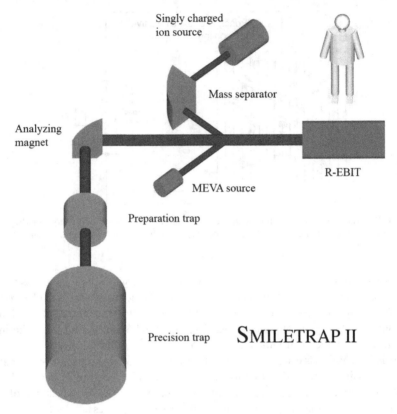

Fig. 12 The Penning-trap mass spectrometer facility SMILETRAP II at the AlbaNova University Center, Stockholm

measurement. The preparation trap is placed in a 1.1 T electromagnet and the precision trap is placed in a 6 T horizontal superconducting magnet.

6.1 Mass Measurement Procedure at SMILETRAP

The SMILETRAP facility [23] consists of a hyperboloid Penning-trap mass spectrometer with a superconducting magnet (Oxford Instruments, NMR Division, Type 200/130, Cryostat Family Type 3) having a 4.7 T central field. The flow of a measurement is illustrated in Fig. 13. The experiment starts at the ion source where the highly charged ions are produced as described in Sect. 3.1. In Fig. 14a, a mass scan is shown where the isotopes of Hg are resolved. A well-defined isotopically pure sample of singly charged ions is injected into CRYSIS for charge breeding. The extracted HCIs are transported to the trap experiment over a distance of about 15 m using conventional ion beam optics including a number of electrostatic deflectors,

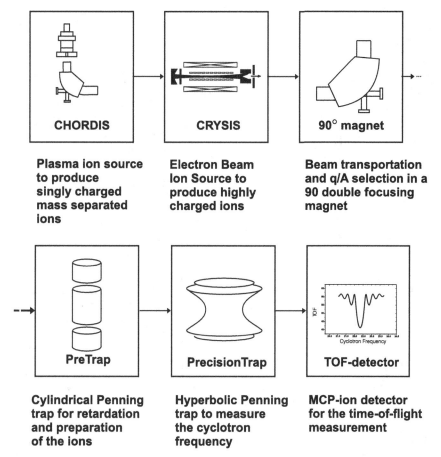

Fig. 13 Schematic illustration of the flow of a typical mass measurement routine with the SMILE-TRAP Penning-trap mass spectrometer

quadrupole triplets, and einzel-lenses. Along the beam line, remotely controlled position-sensitive strip-detectors and Faraday-cups are used to monitor the ion beam profile and to measure the ion beam intensity. The ion beam energy is low, typically $3.4\,keV/q$.

Before entering the first cylindrical Penning trap, named pre-trap, a charge-state selection is done by scanning the current of a 90° double-focusing magnet which has a bending radius of 500 mm. An example of a charge-state spectrum can be seen in Fig. 14b. The same magnet also serves to deflect the beam vertically to the experiment. Out of the 100 μs long CRYSIS pulse, only a small fraction (1–2 μs) is captured in the pre-trap due to the short length of the pre-trap compared to the length of the pulse. Typically, a few thousand ions are captured here. The capturing is done by removing a potential barrier at the entrance side of the trap by applying a short 20 μs negative voltage pulse on the lower end cap of the pre-trap. When optimizing the capture in the pre-trap, the high-voltage level of the pre-trap has to be scanned in

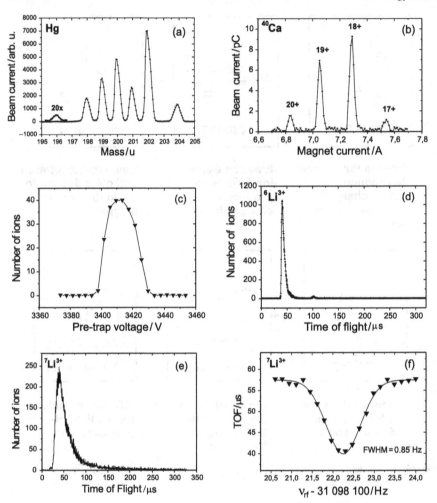

Fig. 14 (**a**) A typical mass spectrum for the mercury isotopes, (**b**) a charge spectrum for ^{40}Ca, (**c**) a pre-trap high-voltage scan, (**d**) a time-of-flight spectrum for ^{6}Li $^{3+}$ ions, (**e**) another time-of-flight spectrum with excited ^{7}Li $^{3+}$ ions, and (**f**) a cyclotron resonance for ^{7}Li $^{3+}$. For further details see text

order to match the beam transport energy, as shown in Fig. 14c. After catching the ions, the potential of the pre-trap is quickly lowered to 0 V in a few milliseconds. This is achieved by using a programmable low-noise high-voltage ramp generator. As a result, the energy of the ions is shifted from $+3.4$ keV $\cdot q$ to 0 keV. The ejection is achieved by removing a potential barrier at the exit side of the pre-trap.

The ions are transferred to the hyperbolic Penning trap, named precision trap, through a series of cylindrical drift tubes biased to -1 kV including a number of variable apertures, electrostatic deflectors, and lenses. A channeltron detector and a MCP detector [88] are used to monitor the beam and help optimizing the injection into the strong magnetic field, where the precision trap is located. These detectors can also be used to measure the flight time of the ions ejected from the pre-trap, see Fig. 14d.

The ions are retarded to ground potential before entering the precision trap. An aperture with a hole of 1 mm in diameter is placed at the trap entrance to prevent ions with too large magnetron radii from entering the precision trap. The capturing in the precision trap is achieved in a similar manner as in the pre-trap: A potential barrier is lowered at the entrance, and before the ions, which have been reflected at the upper end cap, can leave the trap again, the voltage on these electrodes is restored back to the nominal value (approximately 5 V) within $< 1\,\mu s$. The capturing in the precision trap is synchronized to the ejection from the pre-trap; thus, a time-of-flight discrimination can be applied, which means that only ions arriving within a short time window of $< 1\,\mu s$ are captured. Out of the few thousand ions ejected from the pre-trap, typically about 100 ions are captured in the precision trap. For a detailed ion balance, see Table 2. To get rid of the ions with high axial energy, a so-called *boil-off* ion evaporation technique is used. This involves lowering the trapping potential from 5 V to a few mV by ramping up the potential of the ring electrode. After the axially hottest ions have left the trap, the potential is restored. The remaining ions are subjected to an azimuthal quadrupolar radio-frequency excitation.

After excitation, the ions are gently ejected from the trap into the drift section and drift through the fringe field of the magnet to a MCP detector [88] 500 mm upstream. The time of flight to the detector is measured and recorded in a commercial multichannel scaler. A typical time-of-flight cyclotron resonance spectrum obtained with $^7Li^{3+}$ ions is shown in Fig. 14f.

In order to obtain the mass by exploiting (2), in addition to the cyclotron frequency ν_c, the magnetic field value, B, has to be known too. Since the B-field cannot be directly measured with sufficient precision, the cyclotron frequency of a reference ion with well-known mass is measured instead, and the mass of the unknown ion is obtained from the ratio of these frequencies, eliminating the dependence on B. The ideal case would be the use of ^{12}C ions of suitable charge state, because the atomic mass unit is defined as one-twelfth of the mass of a ^{12}C atom, and the electron binding energies are well known. Due to technical reasons, H_2^+ ions were used as reference mass at SMILETRAP since they could be easily produced by electron impact ionization of the residual gas in the pre-trap. In order to minimize unwanted

Table 2 Ion balance showing the approximative ion intensity at different stages of the experiment in a typical measurement. The width of the ion pulse injected into CRYSIS can be from 20 ms to 2 s, depending on the requirements. The width of the extracted pulse is 100 μs

Injected into CRYSIS	50 nC
Extracted pulse (all charge states)	500 pC
Charge-separated pulse	10–20 pC
Captured in pre-trap	2000 ions
Captured in precision trap	100 ions
After evaporative cooling	1–5 ions

effects by possible time-dependent B-field fluctuations, the cyclotron frequency of the main ion and the reference ion is alternately measured within rather short time, approximately 1–2 min.

A measurement results in two interlaced series of frequency scans, one for the ion of interest and one for the reference ion. A scan consists of 21 equidistant frequencies and corresponding flight time values. The data is bunched and a resonance curve for each ion species is generated from 10 to 20 scans. A typical measurement with SMILETRAP involves a few thousand scans corresponding to 20,000 or more ions of each species, requiring a measurement time from 1 day up to 1 week or more. The mass precision in a Penning trap depends directly on the charge state, the magnetic field, the excitation time and the $\sqrt{N}$ of the number of detected ions, see (7).

The recorded time-of-flight data is fitted with the theoretical curve using a least-squares method, and the center frequency is obtained to $\sim 1\%$ of the linewidth [66]. If an excitation time of 1 s is used, this results in a line width ≤ 1 Hz, and thus it is possible to reach a statistical uncertainty of a few parts in 10^{10}, considering a cyclotron frequency ν_c around 36 MHz, typical for HCI in a 4.7 T magnetic field.

The sidebands of the resonances which are related to the conversion of magnetron into cyclotron motion during excitation are suppressed in a typical SMILETRAP resonance, see Fig. 14f. The main reasons are an incomplete conversion of magnetron into cyclotron motion during excitation and the short term instability of the magnetic field.

The center frequencies ν_c and the corresponding uncertainty $\delta\nu_c$ are derived from the fit. The obtained $\nu_{c,\text{main}} \pm \delta\nu_{\text{main}}$ and $\nu_{c,\text{ref}} \pm \delta\nu_{\text{ref}}$ are divided to form a frequency ratio $r_i \pm \sigma_i$. The individual ratios r_i are weighted together to give an average $\overline{R}$ which can be expressed as

$$\overline{R} = \frac{\sum_i r_i \frac{1}{\sigma_i^2}}{\sum_i \frac{1}{\sigma_i^2}}. \tag{8}$$

The weighted average is given together with two errors, σ_{int} and σ_{ext}:

$$\sigma_{\text{int}}^2 = \frac{1}{\sum_i \frac{1}{\sigma_i^2}} \qquad \sigma_{\text{ext}}^2 = \frac{\sum_i \frac{1}{\sigma_i^2}(r_i - \overline{R})^2}{(n-1)\sum_i \frac{1}{\sigma_i^2}}. \tag{9}$$

The internal error σ_{int} is the error of the weighted average of the individual frequency ratio measurements. The external error σ_{ext} is the distribution of the individual measurement r_i around the weighted average $\overline{R}$. In case the distribution of r_i is purely statistical and the uncertainty in the fit of the resonance curve has been calculated correctly, both errors should be equal [89]. In case they differ, that is an indication for possible systematic errors. This does not mean that all systematic errors can be found in that difference.

The mass of the ion is calculated using the weighted average ratio of the measured frequencies:

$$\overline{R} = \frac{v_{\text{main}}}{v_{\text{ref}}} = \frac{q_{\text{main}}}{q_{\text{ref}}} \frac{m_{\text{ref}}}{m_{\text{main}}} . \qquad (10)$$

To obtain the atomic mass M, one has to add to the ion mass given by (10), the mass m_e of the missing q electrons and their total binding energy, see (1):

$$M = \frac{1}{\overline{R}} \frac{q_{\text{main}}}{q_{\text{ref}}} m_{\text{ref}} + q_{\text{main}} m_e - E_B , \qquad (11)$$

where E_B is the total ionization energy which can be calculated [90] or is already available with high precision [91, 92, 93]. The value of the electron mass is $m_e = 5.4857990945(24) \times 10^{-4}$ u with a relative standard uncertainty of 4.4×10^{-10} [94], and therefore, this error contribution to the final uncertainty in the mass can be neglected.

6.2 Challenges

High-precision mass measurements with ion traps using HCI deal with problems common to all high-accuracy experiments. These range from mundane stability problems regarding temperature, pressure, high voltage, magnetic field, beam intensity, or electronic thresholds to the effective suppression of vibrations, electric and magnetic field inhomogeneities, and electronic noise. Excellent vacuum conditions, in some cases to pressures reaching $p < 10^{-12}$ mbar, are mandatory for long storage times of the HCI under investigation and to minimize background effects by collisions with rest gas atoms. Present precision limits of almost all Penning-trap mass spectrometers are the limited temporal magnetic field stability $[(\delta B/\delta t)(1/B) = 10^{-9}/h)]$ and spatial homogeneity ($\pm 1 \times 10^{-8}$ measured over a 10 mm diameter spherical volume) of superconducting magnets with field amplitudes of up to 7 T, and the accurate calibration of the magnetic field amplitudes. Severe limitations in the precision of mass determinations with Penning traps are due to temperature and pressure fluctuations in the helium and nitrogen reservoirs of the superconducting magnets [95, 96, 97]. They cause changes in the magnetic susceptibility of the materials surrounding the trap and thus in the magnetic field homogeneity and strength. Stabilization systems with $\Delta T < 10$ mK and $\Delta p < 0.05$ mbar have been developed and implemented [23, 98, 99].

7 Examples for Applications of High-Precision Atomic/Ionic Masses in Modern Physics

Perhaps the most fundamental and discernible aspect of an atom is its mass, which is due to its relation to the nuclear and atomic binding energies a carrier of unique information. Precise mass measurements have broad spectra of applications in modern physics [25], including new determinations of fundamental constants [100]; a test of the fundamental charge, parity, and time reversal symmetry [101]; a direct test of Einstein's $E = mc^2$ equivalence [102]; understanding astrophysical heavy-element formation [103]; and a possible route to realizing an atomic definition of the kilogram [104, 105]. So far, mass measurements involving ions of about 30 isotopes in the mass range 1–200 u and charges, q from 1+ to 52+, have been performed at SMILETRAP. Among the highlights are the determination of the ^{76}Ge double β-decay Q-value [106] and the mass measurement of ^{133}Cs related to the determination of a new value for the fine-structure constant α [107]. The mass of the ^{28}Si was measured for a possible atomic definition of the kg-standard [108]. By measuring the mass of the ^{198}Hg and ^{204}Hg, a problem in the atomic mass table has been solved [109]. Very recently, by adding a new mass measurement of ^{3}He^{1+} ions, we improved our previous Q-value for the tritium β-decay by a factor of 2, resulting in an uncertainty of 1.2 eV, thus being presently the most accurate tritium decay Q-value, and more importantly, it is based on correct atomic masses [110]. A new mass value for ^{7}Li has been measured, to be used as mass calibration in radioactive beam experiments when measuring the masses of halo nuclei of ^{6}He, ^{8}He, and ^{11}Li [111]. For the evaluation of upcoming g-factor experiments [112], mass values of highly charged ^{24}Mg, ^{26}Mg [113], and ^{40}Ca [114] ions are needed.

In the following, three examples of high-precision mass measurements of highly charged ions will be given with some brief discussion of the physics case.

7.1 The Mass of ^{28}Si and the Definition of the Kilogram SI Unit

The *kilogram* is the SI base unit of mass and is defined as being equal to the mass of the international prototype of the kilogram, see Fig. 15 (left). The international prototype of the kilogram is a cylinder of 39 mm height and diameter made from an alloy of 90% platinum and 10% iridium and it was made in the 1880s. It is kept together with six copies in a vault in air under three bell jars at the Bureau International des Poids et Mesures (International Bureau of Weights and Measures) in Sèvres, France. The kilogram is the only SI base unit that is still defined in relation to a material artifact rather than in terms of some fundamental physical property/constant. Official copies of the prototype kilogram are made available as national prototypes, which are compared to the Paris prototype, *Le Grand Kilo*, roughly every 40 years. The combined standard uncertainty currently assigned to the calibration of a national prototype is typically 5 μg.

The main problems associated with this definition are well known [115, 116] and can be summarized as follows:

- The international prototype can be damaged or even destroyed.
- It is not well defined; it accumulates foreign material and has to be cleaned with unknown effect; a special cleaning procedure was also introduced [117].
- It is probably aging; a significant average drift of about 50 μg of the official and national copies of the international prototype has been observed over 100 years.

Fig. 15 *Left*: The present international prototype of the kilogram kept by the Bureau International des Poids et Mesures in Sèvres near Paris, France (Copyright © BIPM, photo reproduced with the permission of BIPM, France). *Right*: A perfect Si ball being polished, which may be the future kilogram standard. The roundness of the finished sphere is about 50 nm on a 93.6 mm diameter. It is believed to be the roundest and most symmetric object in the world (Copyright © CSIRO, photo reproduced with the permission of CSIRO, Australia)

Because of the number of these secondary standards (about 40) it is rather probable that the international prototype itself is the source of this drift.

• It limits the stability of the kilogram relative to fundamental constants to a few parts in 10^8 at best. It limits also the definition of other SI units, e.g., the ampere, the mole, and the candela, because of the way they are defined [118].

Several possibilities of a new definition are discussed among metrologists today [115, 116, 118, 119]; one of them is to trace back the unit of mass to an atomic mass [116]; another, quite promising, is based on the mass equivalent of energy, the so-called Watt-balance experiment [120].

The kilogram could be defined as the mass of a fixed number of atoms of a definite species [121]. For this, the number of atoms in a macroscopic mass of a substance has to be known. The proposition of the Avogadro Project (laboratories involved are from Germany, Italy, Belgium, Japan, Australia, and the USA) is to redefine the kilogram in terms of the Avogadro constant, N_A. As such, the kilogram could be defined as the mass of $(1000 \cdot N_A)/12$ of ^{12}C atoms. Currently, the Avogadro constant is known only to an uncertainty of approximately 1.7×10^{-7} [94], which should be reduced to less than 1×10^{-8} in the near future. One way toward an improved Avogadro constant is to obtain it from the ratio of the molar mass to the mass of an atom, $N_A = M_{mol}/m_{at}$. Silicon was chosen for this measurement due to its well-known crystal structure, purity, stability, and its relative ease of use. Assuming that the unit cell of the silicon is cubic in shape and has eight atoms, the expression for the Avogadro constant can be written as

$$N_A = \frac{M_{\text{mol}}}{m_{\text{Si}}} = \frac{8vM_{\text{mol}}}{V_0 m}, \tag{12}$$

where M_{mol} represents the mean molar mass of silicon, m is the mass, v is the volume of the silicon sample, and V_0 is the volume of the unit cell.

The volume of the unit cell is $V_0 = a^3$, where a is the lattice spacing and is determined by X-ray interferometry [122, 123]. The atomic masses of the Si isotopes have to be known to a relative mass precision of about 10^{-9} and also the isotopic composition of the crystal. At SMILETRAP, the mass of ^{28}Si has been measured to 3.5×10^{-10} precision using highly charged ^{28}Si^{12+}, ^{28}Si^{13+}, and ^{28}Si^{14+} ions [124]. The MIT group using also Penning traps but with a different technique succeeded in measuring the mass of ^{28}Si with an outstanding uncertainty of only 7×10^{-11} [104]. The two values are in good agreement which gives additional confidence in the two high-precision mass experiments.

For the determination of the lattice spacing, density, and diameter of the sphere, high-precision spheres were manufactured, see Fig. 15 (right). These come in the form of a highly polished 1 kg single-crystal silicon sphere, with a roundness in the range of 50–60 nm. A sphere has been chosen to provide mechanical robustness, no sharp edges or corners which can be easily damaged. The volume is determined from the measurement of the silicon sphere's diameter and roundness. The nominal diameter of a 1 kg Si sphere is 93.6 mm. In order to obtain an accuracy better than 1×10^{-8} in volume, the diameter must be known to a range of 0.6 nm, in other words, within one atom spacing. The limiting factors today are the variability from sample to sample of the isotopic abundances and the content of impurities and vacancies. In addition, these measurements require well-controlled environments since the measurements are sensitive to many parameters, particularly to those of temperature and pressure. An instability within the range of 2 mK would be sufficient to cause the silicon to expand by more than the allowable uncertainty.

It will be difficult to ensure that the silicon sphere remains unchanged, but the major benefit is that with some effort it could be reproduced if damaged.

7.2 Testing Bound-State Quantum Electrodynamics

The g-factor of a particle is a dimensionless constant which determines the strength of the interaction of its intrinsic magnetic moment with a magnetic field. The very accurate measurements of the g-factor of the free electron [125, 126] have been one of the most stringent test of quantum electrodynamics (QED). At the end, for the comparison of experimental g-factor with the QED, the best value for the fine-structure constant α is derived [127]. The success of the QED theory culminated with the excellent agreement of the experimentally measured value of the free-electron g-factor by Dehmelt and Van Dyck [125] and theoretically calculated value by T. Kinoshita [128, 129]. Recently, Gabrielse achieved a six-fold improved relative uncertainty of 0.76×10^{-12} for the free-electron g-factor [126].

In the case of an electron bound to a highly charged ion, the g-factor serves as one of the most stringent tests of bound-state quantum electrodynamics (BS-QED) in the magnetic sector. Quantum electrodynamics is a relativistic quantum field theory of electromagnetism, mostly dealing with particles interacting with each other or with electromagnetic fields by means of exchange of photons [130]. In a hydrogen-like HCI with one electron bound to a nucleus of charge Z, the very strong Coulomb field between the electron and the nucleus leads to relativistic binding effects additionally to QED [131, 132]. So far, the hydrogen-like $^{12}C^{5+}$ [133] and $^{16}O^{7+}$ [134] are the heaviest ions used in the Mainz g-factor experiments, which are compared to theoretical calculations [135, 136]. In these experiments, the value of g is obtained from

$$g = 2\frac{\omega_L}{\omega_c}\frac{q}{e}\frac{m_e}{m_{ion}} , \tag{13}$$

where the spin precession (Larmor) frequency, ω_L, and the true cyclotron frequency, ω_c, are precisely measured [137]; obviously, the ionic mass, m_{ion}, is an essential and necessary ingredient for the evaluation of the g-factor experiments of the electron bound in hydrogen-like and lithium-like ions [138].

The ionic mass m_{ion} is measured directly with SMILETRAP, and thus, there is no need to use theoretically calculated electron binding energies as discussed in [139]. In both g-factor experiments performed so far with highly-charged ions ($^{12}C^{5+}$ and $^{16}O^{7+}$), the ionic masses are accurately known, and by far the dominant contribution to the total uncertainty in g came from the uncertainty in the electron mass m_e. From these measurements combined with theoretical calculations, a new electron mass was derived [140]. The present uncertainty in the electron mass is $\delta m/m = 4.4 \times 10^{-10}$ [94]. The proposed new g-factor experiments involving medium heavy and later heavy ions [138] will benefit from the new high-precision electron mass m_e; thus, these experiments require ion masses with similar uncertainties as the electron mass, see (13). If we consider the ratio of g-factors of hydrogen-like ions of different isotopes of the same element, the dependence on the electron mass can be eliminated, and nuclear contributions can be determined. An isotope effect in the g-factor $\Delta g = g_1/g_2 - 1$ can thus be introduced to characterize these contributions. It was shown that an isotope effect in the g-factor would be measurable already between $^{24}Mg^{11+}$ and $^{26}Mg^{11+}$, thanks to the high-precision mass values measured with SMILETRAP for both isotopes [113].

The next g-factor experiment in Mainz is going to use highly charged Ca ions, which are heavy enough to test the bound-state contributions to the theoretical value a factor of $8\times$ better than the sensitivity achieved previously in [134]. In addition, $^{40}Ca^{19+}$ has a doubly magic nucleus (consisting of 20 neutrons and 20 protons), which results in a better accuracy of the theoretical calculation of the nuclear contribution to the g-factor. The ionization energy of the $1s$ shell of the calcium atom is about 5 keV, which means that it is still possible to create the ions without the need for big facilities, as would be the case for heavier ions. With six stable isotopes, among them the doubly magic ^{48}Ca, which is actually 20% heavier than ^{40}Ca, an isotopic effect [139, 113] in the g-factor could be addressed. For this reason, the mass of ^{40}Ca has been measured with SMILETRAP by using highly charged

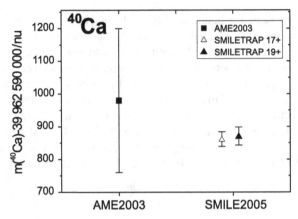

Fig. 16 The mass of ^{40}Ca measured by SMILETRAP [114] compared to previous literature value from [141]

lithium-like (q=17) and hydrogen-like (q=19) ions, and the uncertainty has been improved by a factor of 10 compared to available literature values, see Fig. 16. There is also increasing interest in the properties of lithium-like ions [142] in order to eliminate nuclear size and polarization effects. A comparison of the electronic g-factors of hydrogen-like and lithium-like ions of the same nuclear charge Z can provide a test of bound-state QED in the absence of effects coming from atomic core motion and polarization, since these effects are identical (in first-order approximation) in both species and therefore cancel in a direct comparison. It is thus possible to separate the effects due to the bound state only and benchmark the according QED predictions separately (see [143]). The importance here comes from the fact that for non-light ions the magnitude of nuclear contributions depends strongly on Z. For a review on the importance of g-factor measurements of lithium-like medium to heavy ions see [142]. Within the HITRAP project at GSI, one aims at g-factor measurements of very highly charged ions as, e.g., ^{208}Pb^{81+}, which also has a magic nuclei (82 protons, 126 neutrons) and zero nuclear spin.

7.3 Contributions of Mass Measurements to Neutrino-Physics

The existence of neutrinos was first proposed by Pauli in 1930 [144] in order to explain the continuous energy spectrum of the beta-decay assuming energy conservation is valid. The name *neutrino* was given by E. Fermi in 1933 to distinguish it from the neutron. The experimental proof of their existence came only 26 years later in 1956 by Reines and Cowan [145]. For a brief historical overview, see [146].

There are three known types/flavors of neutrinos called the electron-neutrino (v_e), muon-neutrino (v_μ), and tau-neutrino (v_τ) and their antiparticles. Until recently, according to the Standard Model (SM) of particle physics, neutrinos were assumed to be massless. However, recent investigations of neutrinos from the Sun and of neutrinos created in the atmosphere by cosmic rays have given clear evidence for massive neutrinos indicated by neutrino oscillations [147, 148]. Because oscillation probabilities depend only on squared mass differences, Δm^2, such experiments have no sensitivity to the absolute value of neutrino masses [149]. Neutrinos are extremely important particles in cosmology since they are the second most abundant objects in the universe after photons. The current upper limit on neutrino mass makes neutrinos only a small component of the dark matter [149]. Another implication of neutrino mass is leptogenesis, where different models exist for different neutrino mass regimes [150]. Among the many open questions regarding neutrinos, one is, What mechanism gives neutrinos mass? To understand this better and to answer the many open questions, the mass of the neutrino has to be measured.

A sensitive direct mass measurement is a β-decay spectrum shape measurement. Any isotope that undergoes β-decay could be used for a neutrino mass measurement. Due to a low Q-value (endpoint) and relatively short half-life ($T_{1/2} = 12.32$ years), the tritium β-decay reaction, $^3\text{H} \rightarrow {}^3\text{He} + e^- + \overline{v}_e$, is rather advantageous. Here the tritium (^3H) atom spontaneously emits a single electron (e^-) and an electron-antineutrino ($\overline{v}_e$). The non-zero neutrino mass distorts the shape of the spectrum near the endpoint (see Fig. 17) as predicted by Fermi in 1934 [151]. In the standard β-decay neutrino mass experiments, the deviation from zero-mass value in the Fermi-Kurie plot is used for the determination of the upper limit of neutrino mass. The zero-mass value of the Fermi-Kurie plot with the e-energy abscissa is called the endpoint, which corresponds to the Q-value of the decay, i.e., the mass difference of the ^3H and ^3He atoms. Presently, the most precise value is from the SMILETRAP Penning-trap mass spectrometer with an uncertainty of 1.2 eV [110]. A comparison of the SMILETRAP value to other values from the literature can be seen in Fig. 18. The shape of the β-spectrum will be measured with a projected 0.2 eV sensitivity by the KATRIN experiment [152], which is expected to be able to measure a neutrino mass of 0.35 eV with 5 sigma significance. The precision achieved by SMILETRAP

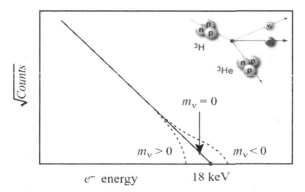

Fig. 17 The last eV region of the β-spectrum of the tritium decay. A non-zero neutrino mass should be observed as a change in the shape of the curve near the endpoint

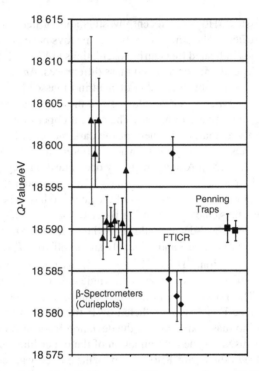

Fig. 18 The Q-value of the tritium β-decay from different experiments [141]. The last value has been measured by SMILETRAP [110]

of δQ=1.2 eV allows only to search for gross systematic errors at KATRIN. For an absolute calibration, even higher precision is necessary to be comparable with the sensitivity of KATRIN. A neutrino is different from the other fermions because it is electrically and color neutral and therefore only interacts weakly. Another interesting unanswered question is whether neutrinos are their own antiparticle (the so-called Majorana particles, after E. Majorana), or do they have separate antiparticles like all known fermions (Dirac particles)? The search for neutrinoless double beta-decay ($0\nu\beta\beta$) tries to answer this question [153], since $0\nu\beta\beta$ can only occur if neutrinos are Majorana particles, because it requires violation of lepton number [154]. Due to the attractive feature that a high-purity germanium (HPGe) single crystal is a combined source and detector, the decay reaction $^{76}Ge \rightarrow {}^{76}Se + 2e + (2\nu)$ has been monitored by the Heidelberg–Moscow collaboration in the Gran Sasso Underground Laboratory. After a data acquisition for about 10 years, the Heidelberg–Moscow collaboration claims that they found evidence for lepton number violation in the decay of ^{76}Ge, at 4.2 σ confidence level [155, 156]. If this is going to be confirmed by other experiments too, this is the first experimental evidence of lepton number non-conservation, giving us an important window on physics beyond the Standard Model. For the evaluation of these experiments, the energy position for a $0\nu\beta\beta$ decay has to be known. As this is a very rare decay and the spectrum is

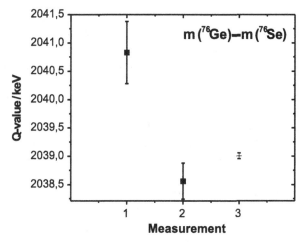

Fig. 19 The Q-value of the ^{76}Ge $\beta\beta$-decay from the previous measurements [141]. The last value has been measured by SMILETRAP [106]

overloaded by background counts, the ^{76}Ge decay Q-value is a needed input, which was provided by SMILETRAP. By performing high-precision mass measurements on highly charged ^{76}Ge and ^{76}Se ions, the Q-value of 2039.006(50) keV was obtained [106], improving the precision of the accepted value by a factor of 6, see Fig. 19.

Recently, the atomic mass of ^{136}Xe was measured by the group of E.G. Myers using a cryogenic Penning trap in a $B = 8.5$ T magnetic field, and multiply charged ions up to ^{136}Xe^{5+}. The measured mass value is $m(^{136}$Xe$) = 135.907\,214\,484\,(11)$ u which if combined with the available mass of ^{136}Ba [141], gives a Q-value 2457.83(37) keV, sufficiently precise for ongoing searches for the neutrinoless double-beta decay of ^{136}Xe [22].

To understand the nature of the neutrino, both $0\nu\beta\beta$ and β-decay experiments are needed, and Penning traps using HCI are providing the necessary precise atomic mass values/mass differences in both cases.

8 Summary and Outlook

Efficient tools for HCI production in the form of electron beam ion traps/sources were developed. They give intense pulses of low-temperature highly charged ions in a very short time interval. These can be re-trapped and prepared by additional cooling to be captured in a precision Penning trap. There the charge merit of HCI is exploited to reach a relative precision in the order of 10^{-10} and in the future even lower to 10^{-11} for any heavy element. Mass measurements were shown to have extensive applications in fundamental physics.

Today, there is no operating Penning-trap mass spectrometer facility using simultaneously the merits of highly charged ions, efficient ion cooling, cryogenic

Penning traps, and non-destructive single ion detection although these techniques are mostly developed in other experiments using Penning traps and HCI. Within the MATS group at the University of Mainz, a new Penning-trap mass spectrometer is being designed, which is going to merge all these techniques and the ultimate goal is to reach ultra-high mass precision in the 10^{-11}–10^{-12} range also for heavy HCI. Penning-trap mass spectrometry of highly charged radioactive ions is foreseen at TITAN at TRIUMF, Canada, and MATS at FAIR, Darmstadt, Germany, where HCI will be produced in an EBIT ion source. The HITRAP project at GSI/Darmstadt is being set up for experiments with slow heavy HCI; here high-precision mass spectrometry is also planned together with other interesting experiments. In Stockholm, Sweden, SMILETRAP II is being built, which is going to be supplied with HCI by the already operational R-EBIT ion source at AlbaNova.

Acknowledgments Sz. Nagy acknowledges financial support by the Alexander von Humboldt Foundation. We gratefully acknowledge support from the Knut and Alice Wallenberg Foundation, the European R&D network HITRAP (HPRI CT 2001 50036), the European network ITSLEIF (026015), and the Swedish Research Council VR. Financial support by the Helmholtz Association of National Research Centres HGF (VH-NG-037) is also acknowledged.

The authors wish to thank all previous and present members of the SMILETRAP group for their excellent work, S. Böhm for operating the R-EBIT source at AlbaNova, and M. Björkhage from the Manne Siegbahn Laboratory for operating the Electron Beam Ion Source CRYSIS with great enthusiasm. The many extra hours and overnight stays are a monumental contribution to the mostly successful mass measurement beam times at SMILETRAP.

References

1. A.H. Wapstra, Phys. Scripta **T59**, 65 (1995).
2. G. Audi, Int. J. Mass. Spectrom. **251**, 85 (2006).
3. J.J. Thomson, Philos. Mag. **44**, 293 (1897).
4. Nobel Lectures, Physics 1901–1921, Elsevier, Amsterdam (1967).
5. F.W. Aston, Nature **105**, 617 (1920).
6. F.W. Aston, Proc. R. Soc. Lond. **115**, 487 (1927).
7. Nobel Lectures, Chemistry 1922–1941, Elsevier, Amsterdam (1966).
8. G. Siuzdak, The Expanding Role of Mass Spectrometry in Biotechnology, MCC Press, San Diego (2003).
9. J.J. Thomson, Philos. Mag. J. Sci. **13**, 561 (1907).
10. A.J. Dempster, Proc. Am. Philos. Soc. **55**, 755 (1935).
11. K.T. Bainbridge and E.B. Jordan, Phys. Rev. **50**, 282296 (1936).
12. J. Mattauch, Phys. Rev. **50**, 617–623 (1936).
13. A.O. Nier and T.R. Roberts, Phys. Rev. **81**, 507–510 (1951).
14. R.C. Barber et al., Rev. Sci. Instrum. **42**, 1 (1971).
15. E. Koets, J. Phys. E: Sci. Instrum. **14**, 12–29 (1981).
16. L.G. Smith and C.C. Damm, Rev. Sci. Instrum. **27**, 638 (1956).
17. L.G. Smith, Phys. Rev. C **4**, 22 (1971).
18. Les Prix Nobel. Almqvist & Wiksell International, The Nobel Foundation (1989).
19. G. Gräff, H. Kalinowsky, and J. Traut, Zeitschrift für Physik A **297**, 35 (1980).
20. R.S. VanDyck, S.L. Zafonte, and P.B. Schwinberg, Hyper. Interact. **132**, 163 (2001).
21. J.K. Thompson, S. Rainville, and D.E. Pritchard, Nature **430**, 58 (2004).

22. M. Redshaw, E. Wingfield, J. McDaniel, and E.G. Myers, Phys. Rev. Lett. **98**, 053003 (2007).
23. I. Bergström et al., Nucl. Instrum. Meth. Phys. Res. A **487**, 618 (2002).
24. K. Blaum et al., J. Phys. B **36**, 921 (2003).
25. K. Blaum, Phys. Rep. **425**, 1 (2006).
26. H.F. Beyer and V.P. Shevelko, Introduction to the Physics of Highly Charged Ions, IOP Publishing, Bristol and Philadelphia (2003).
27. J. Ongena, Phys. Scripta **T123**, 14–23 (2006).
28. F.H. Séguin et al., Rev. Sci. Instrum. **74**, 975995 (2003).
29. W. Greiner, B. Müller, and J. Rafelski, Quantum Electrodynamics of Strong Fields, 2nd edn, Springer, Berlin (1985).
30. B. Franzke, Nucl. Instrum. Meth. **B24/25**, 18 (1987).
31. G. Zschormack et al., Rev. Sci. Instrum. **77**, 03A904 (2006).
32. H. Khodja and J.P. Briand, Phys. Scripta **T71**, 113 (1997).
33. J. Alonso et al., Rev. Sci. Instrum. **77**, 03A901 (2006).
34. V.B. Kutner, Rev. Sci. Instrum. **65**, 1039 (1994).
35. R. Geller, Rev. Sci. Instrum. **69**, 1302 (1998).
36. R. Schuch, In D.C. Lorents, W.E. Meyerhof, and J.R. Petersen, eds., Proceedings of the XIVth International Conference on the Physics of Electronic and Atomic Collisions, Palo Alto, California, USA, July 24–30, 1985, Elsevier Science (1996).
37. R. Schuch et al., J. Phys. B **17**, 2319 (1984).
38. R.D. Deslattes, R. Schuch, and E. Justiniano, Phys. Rev. A **32**, 1911 (1985).
39. F. Herfurth et al., Int. J. Mass Spectrom. **251**, 266 (2006).
40. E.D. Donets, Bull. OIPOTZ **24**, 65 (1969).
41. M.A. Levine, R.E. Marrs, J.E. Henderson, D.A. Knapp, and M.B. Schneider, Phys. Scripta **T22**, 157 (1988).
42. D. Schneider et al., Phys. Rev. A **42**, 3889 (1990).
43. D. Schneider, Hyper. Interact. **99**, 47 (1996).
44. E.D. Donets, Rev. Sci. Phys. **69**, 614 (1998).
45. E. Beebe, L. Liljeby, Å. Engström, and M. Björkhage, Phys. Scripta **47**, 470 (1993).
46. I. Bergström et al., In K. Prelec, ed., Electron Beam Ion Sources and Traps and Their Applications: 8th International Symposium, Upton, New York (2001). AIP Conference Proceedings 572.
47. F. Wenander, Nucl. Phys. A **746**, 40 (2004).
48. E. Kugler, Hyper. Interact. **129**, 23–42 (2000).
49. D. Habs et al., Hyper. Interact. **129**, 43 (2000).
50. B.H. Wolf et al., Nucl. Instrum. Meth. B **204**, 428–432 (2003).
51. J. Dilling et al., Int. J. Mass. Spectrom. **251**, 198 (2006).
52. I.G. Brown, J.E. Galvin, R.A. Macgill, and M.W. West, Nucl. Instrum. Meth. B **43**, 455 (1989).
53. D. Schneider et al., Phys. Rev. A **44**, 3119 (1991).
54. S. Böhm et al., J. Phys.: Conf. Ser. **58**, 303 (2007).
55. R.E. Marrs, S.R. Elliot, and D.A. Knapp, Phys. Rev. Lett. **72**, 4082 (1994).
56. R.E. Marrs, Rev. Sci. Instrum. **67**, 941 (1996).
57. G. Gabrielse, Phys. Rev. A **27**, 2277 (1983).
58. G. Gabrielse and F.C. Mackintosh, Int. J. Mass Spectrom. Ion Process. **57**, 1 (1984).
59. P.K. Ghosh, Ion Traps, Clarendon Press, Oxford (1995).
60. F.G. Major, V.N. Gheorghe, and G. Werth, Charged Particle Traps, Springer, Berlin, (2005).
61. L.S. Brown and G. Gabrielse, Rev. Mod. Phys. **58**, 233 (1986).
62. S. Rainwille, J.K. Thompson, and D.E. Pritchard, Science **303**, 3334 (2004).
63. S. Brunner, T. Engel, A. Schmitt, and G. Werth, Eur. Phys. J. D **15**, 181 (2001).
64. F. Bloch, Physica **19**, 821 (1953).
65. M. Kretzschmar, Phys. Scripta **46**, 544 (1992).
66. M. König, G. Bollen, H.J. Kluge, T. Otto, and J. Szerypo, Int. J. Mass Spectrom. Ion Process. **142**, 95 (1995).

67. R. Ringle, G. Bollen, P. Schury, S. Schwarz, and T. Suna, Int. J. Mass Spectrom. **262**, 3344 (2007).
68. S. Eliseev et al., Int. J. Mass Spectrom. **262**, 4550 (2007).
69. G. Bollen, R. Moore, G. Savard, and H. Stolzenberg, J. Appl. Phys. **68**, 4355 (1990).
70. H.-U. Hasse et al., Int. J. Mass Spectrom. **132**, 181 (1994).
71. A. Kellerbauer, T. Kim, R.B. Moore, and P. Varfalvy, Nucl. Instrum. Meth. A **469**, 276 (2001).
72. G. Zwicknagel, AIP Conf. Proc. **862**, 281 (2006).
73. N. Oshima et al., Nucl. Instrum. Meth. B **235**, 504–508 (2005).
74. L. Gruber, PhD Thesis: Cooling of Highly Charged Ions in a Penning Trap, Technische Universität Graz (2000).
75. L. Gruber et al., Phys. Rev. Lett. **86**, 636 (2001).
76. M. Bussmann, U. Schramm, D. Habs, V.S. Kolhinen, and J. Szerypo, Int. J Mass Spectrom. **251**, 179–189 (2006).
77. J.P. Holder et al., Phys. Scripta **T92**, 158 (2001).
78. G. Gabrielse et al., Phys. Rev. Lett. **63**, 1360 (1989).
79. G. Gabrielse et al., Phys. Rev. Lett. **89**, 213401 (2002).
80. M. Amoretti et al., Nucl. Instrum. Meth. A **518**, 679 (2004).
81. J. Bernard et al., Nucl. Instrum. Meth. **532**, 224228 (2004).
82. H.G. Dehmelt and F.L. Walls, Phys. Rev. Lett. **21**, 127131 (1968).
83. D.J. Wineland and H.G. Dehmelt, J. Appl. Phys. **46**, 919930 (1975).
84. G.I. Budker, Proceedings of International Symposium on Electron and Positron Storage Rings, Saclay, Sept. 2630, Rep. II-1-1 (1966).
85. G.I. Budker, At. Energy **22**, 438 (1967).
86. M. Amoretti et al., Nature **419**, 456 (2002).
87. I. Bergström, In R.S.I. Bergström and C. Carlberg, eds., Proceedings of Nobel Symposium 91 Lysekil, Sweden, August 19–26, 1994, World Scientific, Singapore (1994).
88. J.L. Wiza, Nucl. Instrum. Meth. A **162**, 587 (1979).
89. R.T. Birge, Phys. Rev. **40**, 207 (1932).
90. G.C. Rodrigues, M.A. Ourdane, J. Bieron, P. Indelicato, and E. Lindroth, Phys. Rev. A **63**, 012510 (2000).
91. G.C. Rodrigues, P. Indelicato, J.P. Santos, P. Patté, and F. Parante, At. Data Nucl. Data Tables **86**, 117 (2004).
92. R.L. Kelly, J. Phys. Chem. Ref. Data **16**, Suppl. 1 (1987).
93. J.H. Scofield, LLNL Internal Report UCID-16848 (1975).
94. P.J. Mohr and B.N. Taylor, Rev. Mod. Phys. **77**, 1 (2005).
95. R.S. Van Dyck, Jr., D.L. Farnham, S.L. Zafonte, and P.B. Schwinberg, Rev. Sci. Instrum. **70**, 1665 (1999).
96. G. Gabrielse and J. Tan, J. Appl. Phys. **63**, 5143–5148 (1988).
97. K. Blaum et al., J. Phys. G **31**, 1775 (2005).
98. S. Brunner, T. Engel, and G. Werth, Meas. Sci. Technol. **6**, 222 (1995).
99. M. Marie-Jeanne et al., Nucl. Instrum. Meth. A **587**, 464–473 (2008).
100. M.P. Bradley, J.V. Porto, S. Rainville, J.K. Thompson, and D.E. Pritchard, Phys. Rev. Lett. **83**, 4510 (1999).
101. G. Gabrielse et al., Phys. Rev. Lett. **82**, 3198–3201 (1999).
102. S. Rainville et al., Nature **438**, 1096 (2005).
103. H. Schatz and K. Blaum, Europhys. News **37**, 16–21 (2006).
104. F. DiFilippo, V. Natarajan, K.R. Boyce, and D.E. Pritchard, Phys. Rev. Lett. **73**, 1481 (1994).
105. R. Jertz et al., Phys. Scripta **48**, 399 (1993).
106. G. Douysset, T. Fritioff, C. Carlberg, I. Bergström, and M. Björkhage, Phys. Rev. Lett. **86**, 4259 (2001).
107. C. Carlberg, T. Fritioff, and I. Bergström, Phys. Rev. Lett. **83**, 4506 (1999).
108. I. Bergström, T. Fritioff, R. Schuch, and J. Schönfelder, Phys. Scripta **66**, 1 (2002).
109. T. Fritioff, H. Bluhme, R. Schuch, I. Bergström, and M. Björkhage, Nucl. Phys. A **723**, 3 (2002).

110. Sz. Nagy, T. Fritioff, M. Björkhage, I. Bergström, and R. Schuch, Europhys. Lett. **74**, 404–410 (2006).
111. Sz. Nagy et al., Phys. Rev. Lett. **96**, 163004 (2006).
112. Sz. Nagy et al., J. Phys. Conf. Ser. **58**, 109–112 (2007).
113. I. Bergström et al., Eur. Phys. J. D **22**, 41 (2003).
114. Sz. Nagy et al., Eur. Phys. J. D **39**, 1 (2006).
115. I.M. Mills, P.J. Mohr, T.J. Quinn, B.N. Taylor, and E.R. Williams, Metrologia **42**, 71–80 (2005).
116. P. Becker and M. Gläser, Int. J. Mass Spectrom. **251**, 220 (2006).
117. G. Girard, The procedure for cleaning and washing platinum–iridium kilogram prototypes used at the bureau international des poids et mesures, BIPM Monographie, Sévres (1990).
118. I.M. Mills, P.J. Mohr, T.J. Quinn, B.N. Taylor, and E.R. Williams, Metrologia **43**, 227–246 (2006).
119. J.W.G. Wignal, Meas. Sci. Technol. **16**, 682 (2005).
120. R.L. Steiner, E.R. Williams, D.B. Newell, and R. Liu, Metrologia **42**, 431 (2005).
121. J. Flowers, Science **306**, 1324 (2004).
122. R.D. Deslattes and A. Henins, Phys. Rev. Lett. **31**, 972 (1973).
123. R.D. Deslattes et al., Phys. Rev. Lett. **33**, 463 (1974).
124. I. Bergström, T. Fritioff, R. Schuch, and J. Schönfelder, Phys. Script **66**, 201 (2002).
125. R.S. Van Dyck, Jr., P.B. Schwinberg, and H.G. Dehmelt, Phys. Rev. Lett. **59**, 26 (1987).
126. B. Odom, D. Hanneke, B. D'Urso, and G. Gabrielse, Phys. Rev. Lett. **97**, 030801 (2006).
127. G. Gabrielse, D. Hanneke, T. Kinoshita, M. Nio, and B. Odom, Phys. Rev. Lett. **97**, 030802 (2006).
128. T. Kinoshita, Metrologia **25**, 233 (1988).
129. T. Kinoshita and M. Nio, Phys. Rev. D **73**, 013003 (2006).
130. R.P. Feynman, QED: The Strange Theory of Light and Matter, Princeton University Press (1988).
131. T. Beier et al., Phys. Rev. A **62**, 032510 (2000).
132. T. Beier, Phys. Rep. **339**, 79 (2000).
133. H. Häffner et al., Phys. Rev. Lett. **85**, 5308 (2000).
134. J. Verdú et al., Phys. Rev. Lett. **92**, 093002 (2004).
135. V.A. Yerokhin, A.N. Artemyev, P. Indelicato, and V.M. Shabaev, Nucl. Instrum. Meth. B **205**, 47–56 (2003).
136. K. Pachucki, A. Czarnecki, U.D. Jentschura, and V.A. Yerokhin, Phys. Rev. A **72**, 022108 (2005).
137. G. Werth et al., Int. J. Mass Spectrom. **251**, 152.158 (2006).
138. M. Vogel et al., Nucl. Instrum. Meth. B **235**, 7 (2005).
139. T. Fritioff et al., Int. J. Mass Spectrom. **251**, 281–285 (2006).
140. T. Beier et al., Phys. Rev. Lett. **88**, 011603 (2002).
141. G. Audi, A.H. Wapstra, and C. Thibault, Nucl. Phys. A **729**, 1 (2003).
142. V.M. Shabaev et al., Phys. Rev. A **65**, 062104 (2002).
143. T. Beier, P. Indelicato, V.M. Shabaev, and V.A. Yerokhin, J. Phys. B: At. Mol. Opt. Phys. **36**, 1019–1028 (2003).
144. W. Pauli, *Liebe radioaktive Damen und Herren*, Letter to participants of the Conference in Tübingen (1930).
145. F. Reines and C.L. Cowan, Nature **178**, 446 (1956).
146. P. Ramond, Nucl. Phys. Proc. Suppl. [hep-ph=9809401] **77**, 1 (1999).
147. Y. Fukuda et al., Phys. Rev. Lett. **81**, 1562 (1998).
148. S. Eidelman et al., Phys. Lett. B **592**, 1 (2004).
149. S. Hannestad, Annu. Rev. Nucl. Part. Sci. **56**, 137 (2006).
150. A.D. Dolgov, Phys. Rep. **370**, 333–535 (2002).
151. E. Fermi, Z. Phys. **88**, 161 (1934).
152. J. Angrik et al., FZKA Sci. Rep. **9090**, 1 (2004).

153. S.R. Elliott and P. Vogel, Annu. Rev. Nucl. Part. Sci. **52**, 115–151 (2002).
154. S.R. Elliott and J. Engel, J. Phys. G: Nucl. Part. Phys. **30**, R183R215 (2004).
155. H. Klapdor-Kleingrothaus, A. Dietz, I.V. Krivosheina, and O. Chkvorets, Nucl. Instrum. Meth. A **522**, 371–406 (2004).
156. H. Klapdor-Kleingrothaus, A. Dietz, H.L. Harney, and I.V. Krivosheina, Mod. Phys. Lett. A **16**, 2409 (2001).

Fundamental Tests with Trapped Antiprotons

E. Widmann

1 A Brief History of Antimatter

The history of antimatter goes back to the first part of the twentieth century, when attempts were made to combine the two most modern and revolutionary theories of these days, quantum mechanics and relativity.

The starting point for the search for a relativistically invariant form of the Schrödinger equation was the relativistic energy–momentum relation

$$E^2 - p^2 = m^2 \tag{1}$$

($c = 1$) with $E = T + m$, where E is the total energy, T the kinetic energy, and m the mass of a particle. This led to the Klein–Gordon equation, which is of second order in the time derivative and has the problem that the probability density is not positively definite.

In 1928, Dirac postulated his famous equation for spin-1/2 particles of mass m:

$$(i\gamma^\mu D_\mu - m)\Psi = 0, \tag{2}$$

which is first order in the time derivative and makes use of the Dirac matrices γ^μ ($\mu = 0 \ldots 3$) and spinor wave functions Ψ. Here, $D_\mu \equiv \partial/\partial x_\mu$, x_0 is the time coordinate and $x_1 \ldots x_3$ are the space coordinates. Summation over the four space–time coordinates is implicitly assumed in (2) (for details see any textbook on relativistic quantum mechanics). The Dirac equation is now accepted as the basic equation describing relativistic spin-1/2 particles. It has, however, an intrinsic problem that was immediately realized: because the energy values have to satisfy (1), there exist two solutions for the energy eigenvalues, namely positive and negative ones:

E. Widmann
Stefan Meyer Institute for Subatomic Physics, Austrian Academy of Sciences, Boltzmangasse 3, 1090 Vienna, Austria
e-mail: eberhard.widmann@oeaw.ac.at

Widmann, E.: *Fundamental Tests with Trapped Antiprotons.* Lect. Notes Phys. **749**, 155–188 (2008)
DOI 10.1007/978-3-540-77817-2_6

$$E = \pm\sqrt{p^2 + m^2}. \tag{3}$$

For the interpretation of negative energy solutions, Dirac chose a very bold explanation: he postulated the existence of *antimatter*. Each particle should be accompanied by an antiparticle of same mass and spin but opposite charge. This idea must have been indeed revolutionary at that time, and initially Dirac thought about the proton being the antiparticle of the electron. The situation was solved experimentally when Anderson discovered the positron, the antiparticle of the electron, in cosmic rays in 1932. By then, the concept of antimatter was sound and accepted.

It took more than 20 years for the antiproton to be discovered in an experiment at the Bevatron at Berkeley. Using a proton beam impinging on a stationary target, the threshold energy for the elementary pair production reaction

$$p + p \rightarrow p + p + p + \bar{p} \tag{4}$$

is $T_{th} = 6m_p = 5.63\,\text{GeV}$, which was exactly the maximum energy of the Bevatron. But the protons in the target are part of a nucleus, in which case the threshold energy is slightly reduced due to the Fermi motion ($T_{th}^{nucl} \sim 4.3\,\text{GeV}$), so that enough antiprotons could be produced by the Bevatron.

The first experiment in 1955 by Chamberlain et al. [1] confirmed the production of a particle with negative charge and a mass that was equal to the proton mass within a few percent. The antimatter character of the newly discovered particle was proven by observing annihilations in an emulsion and detecting secondary particles with total kinetic energy exceeding one proton mass [2, 3].

2 Low-Energy Antiproton Facilities – Past, Present, and Future

Dedicated facilities for antiprotons exist at two places in the world, Fermilab (USA) and CERN (Geneva, Switzerland). At Fermilab, only high-energy antiprotons are produced for the Tevatron $p\bar{p}$ collider at TeV energies which among others lead to the discovery of the top quark. A further programme on charmonium spectroscopy was running at the Antiproton Accumulator from 1996 to 2000 (experiment E835 [4]). Only CERN so far has decelerated antiprotons to energies below 1 GeV. Currently, the FAIR facility at Darmstadt is being developed, which will include both a high-energy antiproton ring for charmonium studies and a dedicated low-energy antiproton facility called FLAIR.

2.1 Past: The LEAR Era

Initially, antiprotons were only available as secondary beams like at the Bevatron, i.e. with large emittances and momentum spreads of several percent. Later, in the

1980s, the idea was born at CERN to use proton–antiproton collisions to produce the $W^{\pm}$ and Z_0 intermediate vector bosons predicted by the electro-weak theory. That required the production of antiprotons with high rate and very good beam conditions and thus the AAC, the Antiproton Accumulator Complex, was developed and built at CERN.

Antiprotons were produced from 26 GeV protons provided by the CERN PS (proton synchrotron, cf. Fig. 1). The produced antiprotons have a momentum of about 3.6 GeV/c and are captured in a storage ring. The new feature of the AAC [5] was the accumulation of antiprotons in a storage ring where they can be kept over periods of days. For an efficient capture of a large fraction of $\bar{p}$, it was essential to develop a cooling method to reduce the phase space of the antiprotons captured in a high-acceptance storage ring (AC) so they could be transferred and stored in a dedicated accumulation ring (AA) which had an extremely good vacuum to achieve a long lifetime of $\bar{p}$. The importance of the stochastic cooling technique invented for this purpose can be seen from the fact that Simon Van der Meer received the Nobel Prize for inventing stochastic cooling together with Carlo Rubbia in 1984. In this scheme, antiprotons were produced and "stacked" in the AA every 4 s, leading to a production rate of 10^{12} $\bar{p}$ /day or 1.25×10^7 $\bar{p}$ /s.

Because of the availability of the AAC complex, plans for a low-energy antiproton ring were developed in the early 1980s, and the machine called LEAR (low-energy antiproton ring, see Fig. 2) was built and operated from 1984 to 1996. Antiprotons were extracted from the AA, decelerated by reverse injection into the PS, and transferred to LEAR at 600 MeV/c momentum. In addition to stochastic cooling, LEAR also used electron cooling which is faster at lower energies. Antiprotons could be extracted either continuously ("slow extraction") at a rate of 10^6 $\bar{p}$/s over typically 1 h per fill or pulsed ("fast extraction") in a $\sim$150 ns long pulse containing some 10^9 $\bar{p}$. The momentum (energy) range of LEAR covered 100 MeV/c (5.3 MeV) to 1.96 GeV/c (1.2 GeV).

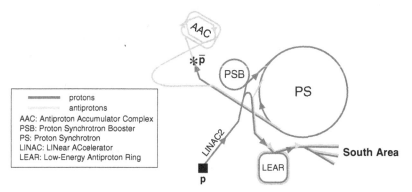

Fig. 1 Accelerator complex of CERN used for antiproton production

Fig. 2 Photograph of the LEAR machine. Clearly visible are the diagonal coaxial copper conductors used for stochastic cooling

2.2 Present: The Antiproton Decelerator (AD) at CERN

With the ending of the high-profile experiment CPLEAR in 1996, CERN decided to shut down LEAR as a contribution to cost reduction in view of the construction of the LHC. But in one of the last experiments at LEAR, a team led by Walter Oelert from Jülich succeeded in observing the first nine atoms of antihydrogen which were created in interactions of antiprotons with an internal gas jet in LEAR [6]. The resulting publicity together with the offer of financial contributions from Japan, Germany, and Italy convinced the CERN management to construct a successor of LEAR, the Antiproton Decelerator (AD) [7] (cf. Fig. 3).

The idea was to create a ring that would combine all the functions that the AC and LEAR had before: capture, cooling, deceleration, and extraction. In such a scheme, no accumulation is possible, thus significantly reducing the production rate of antiprotons. The main reason is the following: the cooling and deceleration of $\bar{p}$ take much longer, namely about 90–100 s, than the 4 s production cycle with the AAC. The whole cycle of the AD is shown in Fig. 4. The resulting antiproton beam is very unique: a pulse of $\sim$2–4 $\times$ 10^7$\bar{p}$ of 100–300 ns length is extracted every 84–100 s. The antiproton energy is restricted to 5.3 MeV (100 MeV/c momentum), although recently higher energy beams have also been extracted for the ACE experiment [8, 9]. This type of beam is well suited for capture in a Penning trap, but cannot be used in nuclear or particle physics-type experiments where single interactions of antiprotons need to be studied.

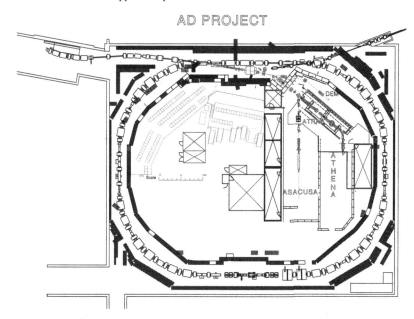

Fig. 3 View of the AD hall at CERN with the three experimental zones inside the ring

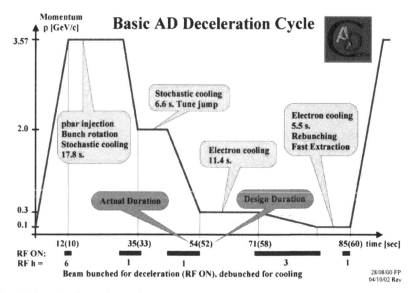

Fig. 4 AD cycle. The red line gives the magnet current and hence the particle momentum as a function of time. The parts where it is constant are the times when cooling is applied. From the capture at 3.6 GeV/c until the extraction at 100 MeV/c it takes 85 s, which is 50% longer than the design value of 60 s shown in brackets

2.3 *Future: The FLAIR Facility*

Within the Facility for Antiproton and Ion Research (FAIR, cf. Fig. 5) at Darmstadt, an antiproton production scheme similar to the AAC of CERN is planned [10]. The initial physics programme behind the antiproton facility is the study of charmonium production and other mesons using a high-energy antiproton storage ring (HESR) and a dedicated detector called $\overline{P}$ANDA (antiProton ANnihilations at DArmstadt) [11]. Soon it was realized that the availability of such an antiproton production scheme would also allow to build a next-generation low-energy antiproton facility. The community submitted a letter of intent for a new facility called FLAIR (Facility for Low-Energy Antiproton and Ion Research) [12] which was soon accepted. FLAIR, as it is planned now, uses the NESR ring of FAIR to decelerate antiprotons to 30 MeV which are then transferred to the FLAIR hall where two storage rings, the LSR (low-energy storage ring, $\overline{p}$ energy 30 MeV – 300 keV) and the USR (ultra-low-energy storage ring, $\overline{p}$ energy 300–20 keV) further decelerate the antiprotons and provide both fast and slow extracted beams in the indicated energy ranges (Fig. 6).

In addition to the two rings, the HITRAP facility, which is currently built at GSI to trap highly charged ions [13], will be moved to FLAIR and can be also used with antiprotons. Furthermore, antiprotons might be extracted from HITRAP and distributed to other experiments at energies around 5 keV, thus creating a second way to generate ultra-low energy $\overline{p}$ beams.

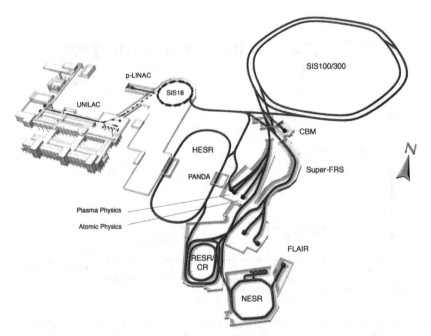

Fig. 5 Overview of the planned FAIR accelerator complex. Antiprotons are captured by the CR and accumulated in the RESR rings. They can then be transferred to the HESR for acceleration or the NESR for deceleration towards FLAIR

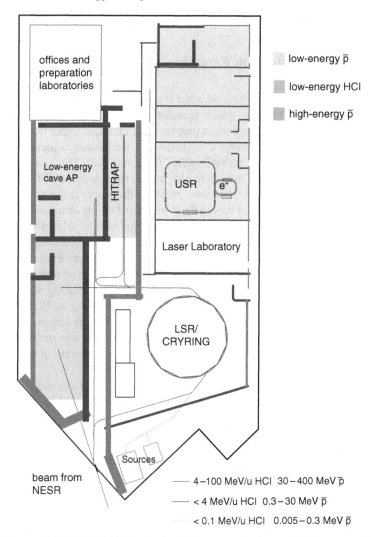

Fig. 6 Layout of the FLAIR hall showing the different areas for low- and high-energy antiprotons and highly charged ions

Due to the space charge limits at low energies in storage rings, FLAIR can make use of only about 10% of the production rate of antiprotons, which makes a parasitic operation with the HESR possible. Nevertheless, due to the high brightness of the cooled $\bar{p}$ beams, rates of stopped antiprotons – in either charged particle traps or low-density gas targets – of 10^6/s are possible, about a factor 100 more than currently available using the 5 MeV AD beam. The availability of slow extraction and the combination with beams of unstable nuclei will make many more experiments possible.

3 $\mathscr{CPT}$ Symmetry and Trapped Antiprotons

3.1 General and Historical Remarks

In the context of studies of antimatter, one immediately encounters the question of matter–antimatter symmetry, which manifests itself in the $\mathscr{CPT}$ theorem: a quantum field theory, which has certain properties, is invariant under the combined operation of the operators $\mathscr{C}$ (charge conjugation, i.e. exchanging a particle with its antiparticle), $\mathscr{P}$ (parity, i.e. mirroring), and $\mathscr{T}$ (time reversal). This is a mathematical theorem which goes back to Lüders [14] and is rigorously valid for quantum field theories formulated on *flat space time*, provided the theory respects (i) *locality*, (ii) *unitarity* (i.e. conservation of probability), and (iii) *Lorentz invariance*.

Among fundamental symmetries, $\mathscr{CPT}$ is unique as it is related to a mathematical theorem based on the above-mentioned properties of quantum field theories. Historically, however, fundamental symmetries played a very important role in the attempt of understanding nature. Initially, physicists believed that all symmetries of space would also be reproduced by nature, and it was a big shock to the scientific community when, after the proposal by Lee and Yang in 1956 [15] that parity might by violated in weak interactions (which they proposed to explain the apparent decay of a K-meson to states of opposite parity), this was really observed 1 year later by C.S. Wu in her famous experiment of the asymmetry in the β-decay of Co [16].

Soon it was seen that these weak interaction processes did conserve $\mathscr{CP}$, i.e. that parity operation followed by an exchange of the particle and its antiparticle. The belief of $\mathscr{CP}$ being conserved held until 1964, when $\mathscr{CP}$ violation was found by Kronin, Fitch, and others in the decay of neutral K-mesons [17]. From that time on, $\mathscr{CP}$ violation was studied in great detail and was so far found only in the weak interaction involving K and recently also B-mesons [18, 19]. A sign of the impact that these findings had on physics is the history of Nobel prizes they were awarded: Dirac, Andersen, Chamberlain and Segre, Lee and Yang, Cronin and Fitch.

The facts that (i) parity is violated to 100% while $\mathscr{CP}$ violation manifests itself by a tiny difference ($\mathscr{O}(10^{-3})$) of certain decay modes and (ii) both types of symmetry violations appear only in the weak interaction, and $\mathscr{CP}$ violations additionally only in the meson sector, have led to speculations that $\mathscr{CPT}$ might also be violated only in a certain sector and to a very small degree.

3.2 Antimatter Absence in the Universe

A strong indication for a distortion in matter–antimatter symmetry comes from the apparent absence of antimatter in the universe (for a detailed discussion on this topic see [20]). Many experiments have searched for antimatter in the universe, but so far no evidence has been found, placing a bound of

$$\beta = \frac{N_B - N_{\bar{B}}}{N_\gamma} \approx 6 \times 10^{-6} \tag{5}$$

on the difference of the cosmological number densities of baryons and antibaryons relative to the photon density [21]. This casts doubt on a scenario where $\mathscr{CPT}$ is conserved during the big-bang.

A solution to this dilemma has been proposed long time ago by Sakharov [22]. In his model, three conditions are necessary to create a baryon excess:

1. Baryon number violation
2. Violation of $\mathscr{C}$ and $\mathscr{CP}$
3. Deviations from thermal equilibrium

The scenario, however, involves $\mathscr{CP}$ violation at high energies which must be different from the observed low-energy $\mathscr{CP}$ violation to quantitatively account for the observed baryon asymmetry [23]. Furthermore, the baryon number violation predicts the decay of the proton which has so far not been observed. Thus, a scenario where $\mathscr{CPT}$ violation occurs in the baryogenesis [24] via the Standard Model Extension of Kostelecky et al. (cf. Sect. 3.3) might offer an alternate explanation for the matter–antimatter asymmetry in the universe.

3.3 Standard Model Extension

In recent years, the group of Kostelecký at Indiana has developed an extension to the standard model that includes both $\mathscr{CPT}$ as well as Lorentz-invariance violating (LIV) terms in the Lagrangian of a quantum field theory [25, 26, 27, 28, 29]. Although this model does not directly predict any $\mathscr{CPT}$ violation nor LIV, it can be used as basis to compare $\mathscr{CPT}$ tests in different sectors and as a guide where to look for possible $\mathscr{CPT}$ violating effects. In fact, various groups have already done so [30, 31, 32, 33, 34, 35, 36].

The modified Dirac equation of SME has the following structure:

$$(i\gamma^\mu D_\mu - m_e - a_\mu^e \gamma^\mu - b_\mu^e \gamma_5 \gamma^\mu - \frac{1}{2}H_{\mu\nu}^e \sigma^{\mu\nu} + ic_{\mu\nu}^e \gamma^\mu D^\nu + id_{\mu\nu}^e \gamma_5 \gamma^\mu D^\nu)\Psi = 0 , \tag{6}$$

where the additional coefficients a_μ^e and b_μ^e violate both Lorentz invariance and $\mathscr{CPT}$, while the other coefficients $H_{\mu\nu}^e$, $c_{\mu\nu}^e$, and $d_{\mu\nu}^e$ violate only Lorentz invariance. The upper index e stands here for the electron, implying that a set of parameters a, B, H, C, d exist for each particle under investigation.

It is important to note that the so-introduced symmetry-violating coefficients have the dimension of mass or energy. This immediately results in the conjuncture that if comparing $\mathscr{CPT}$ tests of different particles or different properties, what counts is the *absolute* precision in eV and not the *relative* precision (cf. Sect. 3.4).

For the case of hydrogen and antihydrogen discussed below, the model [29] has the feature that $\mathscr{CPT}$ violation effects might modify the triplet–singlet hyperfine structures of both hydrogen and antihydrogen, but differently. In the hydrogen atom, this adds an energy correction to states with electron and proton spin components m_J and m_I with value (m_e and m_p denote the electron and proton mass, respectively):

$$\Delta E^{H}(m_J, m_I) = a_0^e + a_0^p - c_{00}^e m_e - c_{00}^p m_p \tag{7}$$
$$+(-b_3^e + d_{30}^e m_e + H_{12}^e)m_J/|m_J|$$
$$+(-b_3^p + d_{30}^p m_p + H_{12}^p)m_I/|m_I| .$$

For antihydrogen, the parameters a, d, and H reverse sign. The anomalous energy terms ΔE^{H} and $\Delta E^{\overline{H}}$ arise from Lorentz invariance violation, among which the parameters a_0s and b_3s are responsible for $\mathscr{CPT}$ violation.

3.4 Status of Verifications of $\mathscr{CPT}$ Symmetry

A summary of the most precise experimental tests of $\mathscr{CPT}$ is shown in Fig.7. Here, two different approaches are shown:

- *Relative precision.* The light bars show particle properties as extracted from the particle data group listings [37]. These numbers are the relative difference of a certain property X measured for a particle and its antiparticle, defined by

$$\Delta_{CPT}(X) = \frac{X(\text{antiparticle}) - X(\text{particle})}{X(\text{particle})} . \tag{8}$$

- *Absolute precision.* Here, the limits directly for the various $\mathscr{CPT}$-violating coefficients of the SME described above are taken from publications. All values

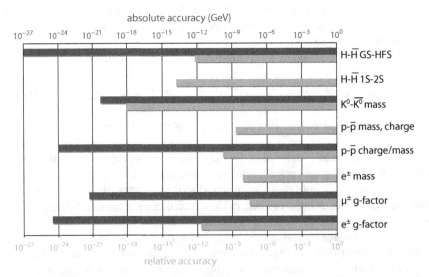

Fig. 7 Comparison of different $\mathscr{CPT}$ tests in terms of absolute and relative precision. The numerical values are presented in Table 1. For details see text

correspond to the current experimental limits except for the case hydrogen–antihydrogen, where it is assumed that $\overline{H}$ could be measured to the same precision as hydrogen has been measured today. For the H–$\overline{H}$ 1S–2S transition, no value is given because it is to first order insensitive to the type of $\mathscr{C}\mathscr{P}\mathscr{T}$ violation appearing in the SME [29].

From Fig. 7, it becomes obvious that when comparing different $\mathscr{C}\mathscr{P}\mathscr{T}$ tests based on absolute accuracy, the low-energy atomic physics-type measurements promise much higher precision than the usually quoted mass difference of $K^0 - \overline{K^0}$ which has the highest relative precision. This approach is valid even if the SME is not considered as the underlying model, since the strength of $\mathscr{C}\mathscr{P}\mathscr{T}$ violation must express itself in a parameter in the Lagrangian of a field theory, which essentially requires the use of an energy scale.

A second point to make is the fact that the $\mathscr{C}\mathscr{P}\mathscr{T}$ tests listed in Fig. 7 involve a great variety of particles, from leptons, mesons, baryons even to atomic systems. As stated above, since, e.g. $\mathscr{C}\mathscr{P}$ violation is so far restricted to the weak interaction in the meson sector, and no prediction exists, where $\mathscr{C}\mathscr{P}\mathscr{T}$ violation might occur preferentially, all sectors need to be investigated with the highest precision possible.

4 Experimental Tests of CPT Using Antiprotons

Antiprotons were first trapped in a Penning trap by the TRAP collaboration led by Gabrielse at LEAR in 1986 [38], for the purpose of comparing the proton and antiproton cyclotron frequency, i.e. their charge-to-mass ratio. In addition, there are two types of experiments underway or planned with the goal of testing $\mathscr{C}\mathscr{P}\mathscr{T}$ with antiprotons: (i) the precision spectroscopy of *antiprotonic helium*, an exotic three-body system $\overline{p}$–e^-–$^4He^{2+}$ which has metastable states where the $\overline{p}$ remains "trapped" with an average lifetime of $\tau \sim 3$ µs. Antiprotonic helium was first discovered at KEK, Japan, in 1991 [39] and later studied at LEAR by the PS205 and at the AD by the ASACUSA collaborations. (ii) *Antihydrogen*, the simplest antimatter atom consisting of a positron and an antiproton. The first nine relativistic atoms of $\overline{H}$ were produced in the last year of LEAR in 1996 [6]. Later experiments at Fermilab [40] confirmed the production of $\overline{H}$ in this reaction with much larger statistics but found a production cross-section that is an order of magnitude smaller than that assumed in the LEAR experiment. The latter two experiments are now being actively pursued at the AD of CERN by the ATRAP, ATHENA/ALPHA, and ASACUSA collaborations.

4.1 Cyclotron Frequency Measurement of the Antiproton

The cyclotron frequency ω_c of a particle is given by its charge Q, mass M, and the external magnetic field B as

$$\omega_c = \frac{Q}{M}B.\qquad(9)$$

The quest was to measure ω_c for both proton and antiproton in exactly the same magnetic field in a Penning trap because the accuracy of the value of B directly enters the error on ω_c or on Q/M.

The first step was to trap $\bar{p}$ in a Penning trap. Usually, the trap electrodes are of hyperbolic shape and the ions are created by ionizing a gas inside the trap by an electron beam or laser. Since $\bar{p}$ are produced at high energies and leave the storage rings at typically 5 MeV kinetic energy, a different design of the trap had to be developed. Figure 8 shows schematically the penning trap as it was used for the first time in this experiment: the electrodes consist of cylindrical rings that leave enough space for injecting particles along the cylinder axis (which is parallel to the external magnetic field).

The parabolic dependency of the confining potential along the cylinder axis was created by an appropriate choice of the length and potential of the different electrodes. The trap is loaded the following way:

1. Antiprotons are extracted in a short pulse (length about 100 ns) and degraded from 5 MeV kinetic energy to <5 keV by this degrader. This method is very inefficient (typically $\sim 10^{-4}$), but precision measurements require ideally only one antiproton in the trap.
2. The opposite side electrode from the $\bar{p}$ entrance is put to a few keV potential (U_5 in Fig. 8), while the entrance electrode is grounded ($U_1 = 0$).
3. The antiprotons are reflected from the downstreams electrode and fly back towards the entrance.
4. In the meantime, the potential is raised on the entrance electrode ($U_1 = U_5$) so that the $\bar{p}$ are also reflected from there and trapped inside.

The max. $\bar{p}$ energy for trapping is determined by the few 100 ns time needed to switch the potential from the entrance electrode from 0 to a value of several keV. The antiprotons are then cooled by collisions with pre-loaded electrons,

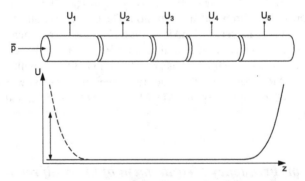

Fig. 8 Schematic view of a cylindrical Penning trap. U_1–U_5 are electrical potentials supplied to the different electrodes to create the trapping potential. During loading with $\bar{p}$, U_1 is set to zero so that particles can enter. Afterwards U_1 is raised to close the trap

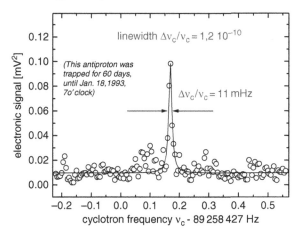

Fig. 9 Measurement of the cyclotron frequency of one trapped antiproton by the TRAP collaboration

which themselves cool through synchrotron radiation in the magnetic fields of a few Tesla strength. This method is still today used by three collaborations, ATRAP ATHENA/ALPHA, and ASACUSA at the AD.

In a series of measurements, the cyclotron frequency of proton and antiproton was measured to higher and higher accuracy by alternately loading the same trap with protons and antiprotons. Figure 9 shows one result of a frequency measurement with a line width of about 10^{-10}. The most precise measurement then used a simultaneously trapped $\bar{\text{p}}$ and a negative hydrogen ion, H^- [30]. The result was a comparison of the charge-to-mass ratio at a level of 10^{-10}:

$$\frac{Q}{M}(\bar{\text{p}})/\frac{Q}{M}(\text{p}) + 1 = 0.9(9) \times 10^{-10}, \tag{10}$$

which is still the most precise test of $\mathscr{CPT}$ invariance in the baryon sector today. The value was slightly modified in 2004, when a polarization effect of molecular ions in Penning traps was discovered [41]:

$$\frac{Q}{M}(\bar{\text{p}})/\frac{Q}{M}(\text{p}) + 1 = 1.6(9) \times 10^{-10}, \tag{11}$$

which is about one sigma off from the $\mathscr{CPT}$ value.

4.2 Precision Spectroscopy of Antiprotonic Helium

4.2.1 General Features

Antiprotonic helium, in short $\bar{\text{p}}\text{He}^+$, is a neutral exotic three-body system consisting of an antiproton, an electron, and a helium nucleus (cf. Fig. 10, left) that has

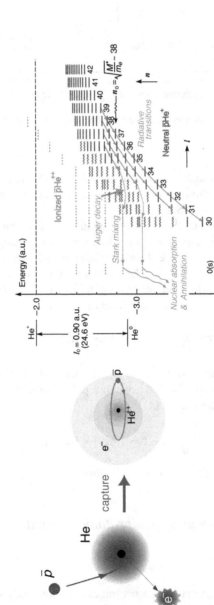

Fig. 10 *Left*: Formation of antiprotonic helium. Because of the resemblance of the exotic three-body system with both an atom and a molecule (due to the large mass of the antiproton), it is often called an "atomcule." *Right*: Energy level diagram

been called a "naturally occurring trap for antiprotons." It is formed when antiprotons are stopped in helium medium and the antiproton kinetic energy falls below the ionization energy of helium (24.6 eV). As shown in Fig. 10, right, the antiproton (like any other negatively charged particle forming an exotic atom) is captured in a principal quantum number $n_0 \sim \sqrt{M^*/m_e}$, where M^* is the reduced mass of the system and m_e is the electron mass. The states in this region are divided into two types: *metastable* (solid lines), where radiative transitions dominate the further de-excitation of $\bar{p}He^+$, and *short lived* (wavy lines). For the latter, Auger decay is the dominating process, leading to lifetimes of 10 ns or less.

The resulting $\bar{p}He^{2+}$ ion is then rapidly destroyed in collisions with surrounding helium atoms through Starck-mixing to states with low angular momentum quantum number l which have a large overlap with the nucleus and from where immediate annihilation of the $\bar{p}$ with one of the nucleons follows. Figure 11 shows the way antiprotonic helium was discovered at KEK in 1991 [39]: by stopping antiprotons from the KEK-PS in liquid helium and measuring the time between a $\bar{p}$ enters and its annihilation into pions. The resulting time spectrum (Fig. 11, right) has two components: 97% of stopped $\bar{p}$ annihilate within nanoseconds, while $\sim$3% survive with an average lifetime of $\sim$3 μs.

Initially, many experiments were done by just measuring the "annihilation" time of antiprotons in various phases of helium and helium plus admixtures of other noble gases or diatomic molecules at LEAR in 1991–1993. A large amount of data on atomic and atomic collision physics was thus collected [42, 43].

4.2.2 Laser Spectroscopy: Charge and Mass of the Antiproton

The fact that the $\bar{p}$ survives for several microseconds in $\bar{p}He^+$ while undergoing radiative transitions whose wavelengths lie in the visible region made it well suited for laser spectroscopy. The technique developed called "forced annihilation" made

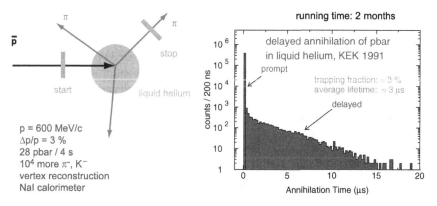

Fig. 11 *Left*: Schematic layout of the first experiment discovering antiprotonic helium. *Right*: Time spectrum (semi-logarithmic scale) of the survival time of antiprotons in liquid helium

use of the additional fact that antiprotons mostly follow cascades with $\Delta n = \Delta l = -1$ or $\Delta v = 0$, with $v = n - l - 1$ being the vibrational quantum number. At the end of each such cascade of Fig. 10, right (see also blow-up in Fig. 12, left), there is a pair of adjacent metastable and short-lived states. When the laser is on resonance, the antiproton is transferred to the short-lived state and annihilates immediately, leading to a sharp spike in the DATS (delayed annihilation time spectrum). Figure 10, right, shows the first such resonance found in 1993 [44].

At the end of LEAR, a precision of 0.5 ppm was obtained for the experimental value of two laser transitions [45]. Comparison with theoretical calculations of the structure of an isolated atomcule showed two important effects (cf Fig. 13, left): at a level of 50 ppm, relativistic corrections due to the movement of the electron have to be included [46], and at about 10 ppm, the first QED effect, the classical Lamb shift, appears [50].

At the AD, the experimental accuracy has been steadily improved. A first major improvement came from the construction of a radio frequency quadrupole decelerator, RFQD, in a joint venture of the ASACUSA collaboration and the CERN PS division [54]. This RFQD decelerates the AD beam from 5 MeV to 60–120 keV, thus allowing to stop antiprotons in helium gas of three orders of magnitude less density. This way, the dominating systematic error of density shifts of the transition wavelengths could be avoided. By comparing the experimental results with two independent theoretical calculations, a $\mathscr{C}\mathscr{P}\mathscr{T}$ limit of 10 ppb on the maximum relative difference of charge and mass of the antiproton could be established (see Fig. 13, right [51]) as discussed below.

The next step in accuracy came with utilizing pulse-amplified cw lasers that were locked to a frequency comb. This way, the two now dominant sources of systematic errors, the laser band width and the absolute calibration of its wavelength, could be strongly reduced. Again the same 13 transitions were measured and compared to the remaining theory that agrees well with the data. By taking the antiproton-to-electron mass ratio $M_{\bar{p}}/m_e$ as a free parameter in the theory and fitting all data points for the various transitions, a value of

$$M_{\bar{p}}/m_e = 1836.152\,674\,(5) \tag{12}$$

was deduced, which is in good agreement with the latest proton value from CO-DATA (cf. Fig. 14, left) [55]. The error on this figure is about 3 ppb.

This treatment assumes that the charge of proton and antiproton has the same value of $|e|$. However, as pointed out by Hughes and Deutch earlier [56], the independent limits on the charge of p and $\bar{p}$ are much less, about $|Q_p - Q_{\bar{p}}|/e < 2 \times 10^{-5}$. In order to extract independent limits for both charge and mass, one can use the following way: The TRAP collaboration has measured Q/M to high precision. The measured transition wavelengths of $\bar{p}He^+$ are dominated by the Rydberg constant which is $\propto MQ^2$. As in [42, 45], the dependence of the measured transition wavelengths of Q and M is as follows: From $v \propto MQ^2$ results a relation for the deviations

$$\frac{\delta v}{v} = \frac{\delta M}{M} + 2\frac{\delta Q}{Q}, \tag{13}$$

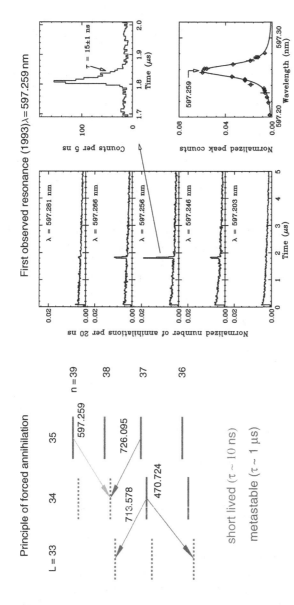

Fig. 12 *Left*: Principle of forced annihilation. *Right*: First 5 μs of DATS with changing laser wavelength. On resonance a sharp spike appears at the time of the laser shot (*upper right*). Plotting the relative area of this spike relative to the total DATS area vs. the frequency allows to determine the central frequency of the transition (*lower right*)

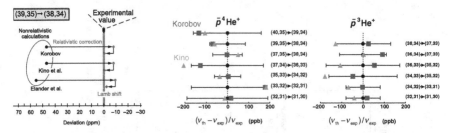

Fig. 13 Comparison of theory and experiment: *Left*: At a level of 0.5 ppm experimental accuracy for one transition [45]. Theoretical results from Korobov [46, 47], Kino [48, 49], and Elander [50] are shown. *Right*: At 10 ppb for a total of 13 transition in both $\bar{p}^4He^+$ and $\bar{p}^3He^+$ [51]. Here newer calculations of Korobov [52] and Kino [53] are compared to experiment

with $\delta M = M_p - M_{\bar{p}}$ and $\delta Q = Q_{\bar{p}} + Q_p$. Since the trap results imply $\delta M/M = \delta Q/Q$,

$$\frac{\delta v}{v} = 3\frac{\delta M}{M} = 3\frac{\delta Q}{Q}. \tag{14}$$

Figure 14, right, illustrates this relation: the allowed region for charge and mass of the antiproton is where the two areas of opposite-sign slope intersect.

In general, each measured transition frequency depends in a slightly different way on charge and mass than in the simple model used above. The relation becomes

$$\delta_p = \frac{Q_{\bar{p}} + Q_p}{Q_p} \sim \frac{M_p - M_{\bar{p}}}{M_p} = \frac{1}{f}\frac{v_{th} - v_{exp}}{v_{exp}}, \tag{15}$$

with $f = 2$–5 being a transition-dependent "sensitivity factor" obtained by theory. The final result is obtained by averaging over all measured transitions. Since the TRAP measurement of Q/M is still one order of magnitude more precise the $\bar{p}He^+$ spectroscopy, the errors on the individual properties are given by the ASACUSA error of 2 ppb [55].

A further improvement from this point is only possible by using two-phtoton spectroscopy, since the line width is now dominated by Doppler broadening. A first test was done in 2006 and the results look encouraging.

4.2.3 Laser-Microwave-Laser Spectroscopy: Magnetic Moment of the Antiproton

The energy levels in antiprotonic helium which were so far denoted by principle n and angular momentum quantum number l exhibit a hyperfine (HF) structure due to the magnetic interaction of the constituents of $\bar{p}He^+$. Since the electron in $\bar{p}He^+$ is predominantly in the ground state, the total angular momentum (l) is equal to the one of the antiproton, $L_{\bar{p}}$. The magnetic moment of $\bar{p}$ is given by the sum of angular and spin terms

$$\mu_{\bar{p}} = (g_\ell^{\bar{p}}L_{\bar{p}} + g_s^{\bar{p}}S_{\bar{p}})\mu_{\bar{N}}, \tag{16}$$

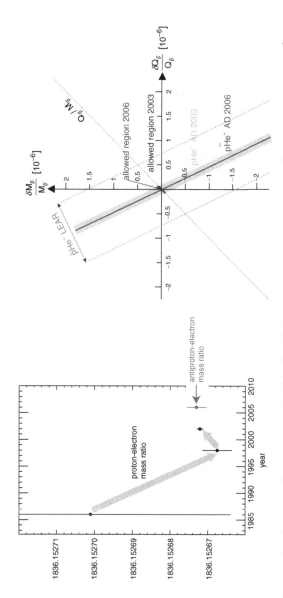

Fig. 14 *Left:* Proton and antiproton-to-electron mass ratio. *Right:* Combining two measurements of TRAP (Q/M) and ASACUSA (MQ^2) to extract individual limits on antiproton and proton charge and mass

where $\mu_{\overline{N}} = Q_{\overline{p}}\hbar/(2M_{\overline{p}})$ is the antinuclear magneton and $g_{\ell}^{\overline{p}}$ and $g_s^{\overline{p}}$ are the orbital and spin g-factors of the antiproton, respectively. Evidently, $g_{\ell}^{\overline{p}}$ is expected to be equal to 1, but this has never been measured for neither proton nor antiproton bound to an atom.

The electron magnetic moment is simply given by its spin part

$$\boldsymbol{\mu_e} = g_{\ell}^e \boldsymbol{L_e} , \tag{17}$$

and the interaction of the two magnetic moments leads to an unusual splitting: the two largest moments, the $\overline{p}$ *angular* moment and the *electron* spin create the dominant splitting following $\boldsymbol{F} = \boldsymbol{Lp} + \boldsymbol{Se}$ called *hyperfine* (HF) splitting, and the $\overline{p}$ spin leads to a further splitting according to the total angular momentum $\boldsymbol{J} = \boldsymbol{F} + \boldsymbol{Sp} = \boldsymbol{Lp} + \boldsymbol{Se} + \boldsymbol{Sp}$. The latter is called *superhyperfine* splitting (SHF), and the resulting quadruplet structure is shown in Fig. 15, left.

The hyperfine structure of $\overline{p}He^+$ has first been calculated by Bakalov and Korobov [58], who showed that the HF splitting is in the order of $\nu_{HF}= 10$–15 GHz for the metastable states, while the SHF splitting is two orders of magnitude smaller ($\nu_{SHF} = 100$–300 MHz). They furthermore showed that the difference of hyperfine splittings in laser transitions is very small for favoured transitions with $\Delta\nu = 0$ (so called because of the larger dipole transition moment), but exceeds the experimental resolution of ~ 1 GHz in unfavoured $\Delta\nu = 2$ transitions. A scan of the $(n,L) = (37,35) \rightarrow (38,34)$ transition done in the last year of LEAR indeed revealed a doublet structure with a splitting of $\Delta\nu_{HF}= (1.70 \pm 0.5)$ GHz [57] (cf. Fig. 15, right), in accordance with the theoretical value of 1.77 GHz [58].

In this experiment, the *difference* of the HF splittings of the two states $(37,35)$ and $(38,34)$ was measured to about 3% precision. In order to directly observe HF transitions within one state (n,L) and to determine the HF splitting to much higher precision, we devised a laser-microwave-laser resonance method which works as follows: The wavy lines in Fig. 15, left, represent allowed M1 transitions (flipping S_e but not $S_{\overline{p}}$) which can be induced by microwave radiation. All the HF levels are initially nearly equally populated. In order to create a population asymmetry which is needed to detect a microwave transition, a laser pulse stimulating a transition from a metastable ($\tau \sim \mu s$) state to a short-lived ($\tau \lesssim 10\,ns$) state can be used. When the $\overline{p}$ is excited to the short-lived state, the $\overline{p}He^+$ undergoes an Auger transition to a $\overline{p}He^{2+}$ ion which is immediately destroyed via collisional Stark effect in the dense helium medium followed by annihilation of the $\overline{p}$ with a nucleon. An on-resonance laser pulse therefore superposes a sharp spike onto the analog delayed annihilation spectrum (ADATS) (cf. Fig. 16A) whose area is proportional to the population of the metastable state at the time of the arrival of the laser pulse.

The laser-microwave-laser resonance method utilizes the following sequence: (i) a laser pulse tuned to one of the doublet lines (e.g. f_+ in Fig. 15, left) preferentially depopulates the F^+ over the F^- doublet. (ii) The microwave pulse is applied; if it is resonant with either ν_{HF}^+ or ν_{HF}^-, it transfers population from the F^- to the F^+ doublet. (iii) A second laser pulse at frequency f_+ measures the new population of F^+ after the microwave pulse. Thus, when plotting the ratio of intensities of the

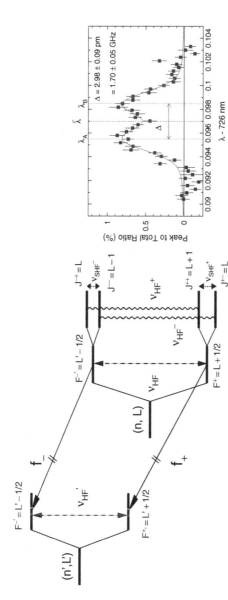

Fig. 15 *Left:* HF splitting of a state (n, L) and observable laser transitions to a daughter state (n', L'). *Right:* First observation of a doublet splitting in a laser transition [57] exhibiting the difference of the f_+ and f_- transition from the HF states. The transitions from the SHF states cannot be observed because of the laser band width and Doppler broadening

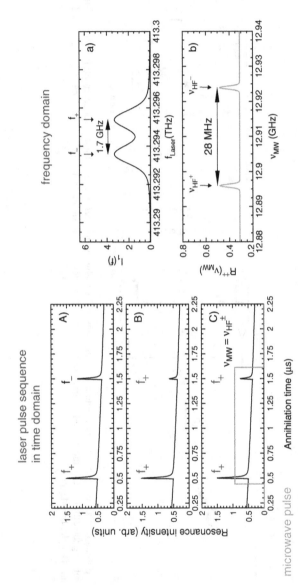

Fig. 16 *Left*: ADATS with a sequence of two laser pulses tuned to different HF lines (A) or the same line (B), where the second pulse has a lower intensity due to the population asymmetry created by the first pulse. (C) The effect of an on-resonance microwave pulse applied between the two laser pulses is shown: the population of the F^+ state is enhance due to the resonant transfer, and the intensity of the second laser pulse is enlarged. *Right*: (a) simulated laser scan and (b) microwave scan results

laser-induced peaks $R^{++} = I_+(t_2)/I_+(t_1)$ as a function of microwave frequency, two peaks should appear as v_{HF}^+ and v_{HF}^- as shown in Fig. 16b).

The experiment has been performed at the AD, and the result indeed showed the expected two resonance lines (cf. Fig. 17, left [59]). The two lines were measured with a relative accuracy of $\sim 3 \times 10^{-5}$ and agreed with most theoretical calculations at a level of $\sim 6 \times 10^{-5}$, which is comparable to the expected accuracy of the calculations originating from the omission of terms of relative order $\alpha^2 \approx 5 \times 10^{-5}$ (cf. Fig. 17, right). The width of the resonance lines of (5.3 ± 0.7) MHz is given by the Fourier limit caused by the time distance of $\Delta t = t_2 - t_1 = 160$ ns between the two laser pulses, leading to an expected width of $\Delta v_{MW} = 1/\Delta t = 6.25$ MHz.

The agreement between theory and experiment can be used to constrain the values of the magnetic moment of the antiproton. Since the observed transitions $v_{HF}^{\pm}$ involve a spin flip of the electron, they are primarily sensitive to the orbital magnetic moment of $\bar{p}$, i.e. the orbital g-factor $g_\ell^{\bar{p}}$. The results imply $|g_\ell^{\bar{p}} - 1| < 6 \times 10^{-5}$. This result is unique since no corresponding value for the proton has ever been measured due to the absence of protonic atoms in our matter-dominated world.

The spin magnetic moment $\boldsymbol{\mu}_{\bar{p}}$ (or equivalently $g_s^{\bar{p}}$) is of greater importance for tests of $\mathscr{CPT}$, since it is a direct property of the antiproton, and its value so far is known only to a precision of $\sim 0.3\%$ [37] from the measurement of X-rays of antiprotonic lead [63]. As recently shown by Bakalov and Widmann [64], the sensitivity of the $v_{HF}^{\pm}$ transitions on $\boldsymbol{\mu}_{\bar{p}}$ is rather small and their measurement does not promise an improvement of the current PDG value. The difference $\Delta v_{HF} = v_{HF}^- - v_{HF}^+$, however, is equal to the difference of SHF splittings $v_{SHF}^+ - v_{SHF}^-$ and is therefore directly sensitive to $\boldsymbol{\mu}_{\bar{p}}$. The experimental error in Δv_{HF} is much larger than the one of $v_{HF}^{\pm}$, and using a sensitivity factor from [64], the current experimental precision corresponds to an uncertainty in $\boldsymbol{\mu}_{\bar{p}}$ of $\sim 1.6\%$.

An improvement of the experimental precision is only possible if the line width can be reduced, i.e. the laser pulse distance Δt can be prolonged. Using the new pulse-amplified cw-laser developed by ASACUSA in 2004, first tests were performed in 2006 showing that this is indeed possible. An improvement of the experimental precision of an order of magnitude would be expected, which should allow to determine $\boldsymbol{\mu}_{\bar{p}}$ to 0.1% or better.

4.3 Precision Spectroscopy of Antihydrogen

4.3.1 General Features

As an ideal "laboratory" to study $\mathscr{CPT}$, antihydrogen has been discussed for a long time (see, e.g. [65]). The main motivation for this is the fact that the $\mathscr{CPT}$ conjugate system, hydrogen (cf. Fig. 18) has been investigated experimentally with higher and higher precision for about 100 years. Indeed, two of the physical quantities that have been measured with the highest precision are the hydrogen 1S–2S two-phtoton transition (relative precision $\sim 10^{-14}$) and the hydrogen ground-state hyperfine structure (relative precision $\sim 10^{-12}$, see Fig. 19 and Table 1).

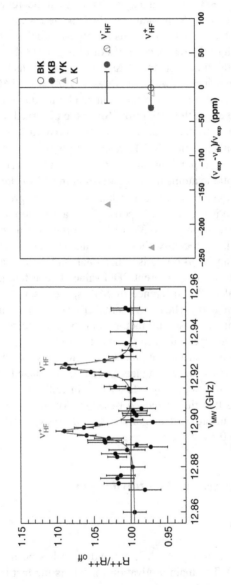

Fig. 17 *Left*: Experimental result of the laser-microwave-laser experiment [59]. *Right*: Comparison of the experimental result with different calculations. The relative deviation of theory and experiment in ppm is plotted [58, 60, 61], K: [62]

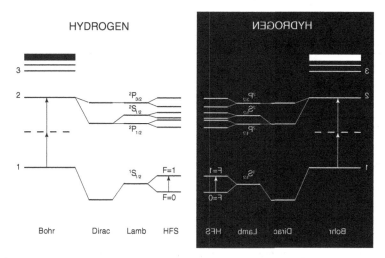

Fig. 18 Hydrogen and antihydrogen energy levels exhibiting a sequence of theoretical models used for their description

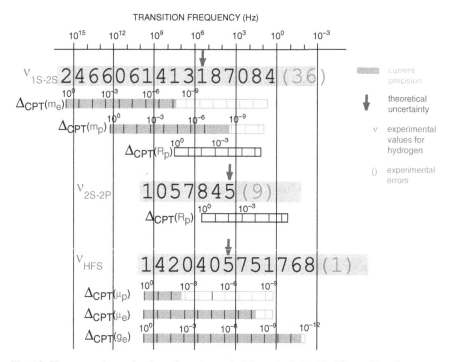

Fig. 19 Three experimental values (large numerical letters) of the 1S–2S transition frequency, 2S–2P Lamb shift, and the 1S hyperfine frequency of hydrogen are presented together with the theoretical uncertainties. Known information on CPT symmetry is also shown (cf. Table 1)

Table 1 Measured quantities of proton, electron, and hydrogen

Measured quantities of hydrogen					
Quantity	Exp. value (Hz)	δ_{exp}/ν	Ref.	δ_{th}/ν	Ref.
ν_{1S-2S}	2,466,061,413,187,084(36)[a]	1.4×10^{-14}	[66]	1×10^{-11}	[67, 68]
ν_{2S-2P}	$1,057,845(9) \times 10^3$	8.5×10^{-7}	[69]	8×10^{-6}	[70]
ν_{HFS}	1,420,405,751.768(1)	7.0×10^{-13}	[71]	$(3.5 \pm 0.9) \times 10^{-6}$ [b]	[72]

Measured CPT quantities as defined in (8)		
Quantity	Value	Ref.
$\Delta_{CPT}(m_e)$	8×10^{-9}	[73]
$\Delta_{CPT}(m_p)$	2×10^{-9}	[55]
$\Delta_{CPT}(R_p)$	—	
$\Delta_{CPT}(\mu_p)$	$(-2.6 \pm 2.9) \times 10^{-3}$	[74]
$\Delta_{CPT}(\mu_e)$	8×10^{-9c}	
$\Delta_{CPT}(g_e)$	$(-0.5 \pm 2.1) \times 10^{-12}$	[75].

m_e: electron mass, m_p: proton mass, R_p: proton radius, μ_e: electron magnetic moment, μ_p proton magnetic moment, g_e electron g -factor. δ_{exp} and δ_{th} denote the experimental or theoretical errors
[a] This is the value for the hyperfine centroid and is obtained by adding a correction for the hyperfine splitting to the value shown in Fig. 2 of Fischer et al. [66] as described in an earlier publication [76]. The quoted error is the quadratic sum of the statistical error shown in the above-mentioned figure, the systematic error of 23 Hz, and the error of the HF correction of 13 Hz [76]
[b] Difference between theory and experiment $(\nu_{th} - \nu_{exp})/\nu_{exp}$
[c] The accuracy of the electron magnetic moment is determined by the accuracy of the electron mass

Figure 19 compares the two already mentioned quantities of hydrogen plus the classical Lamb shift, together with the current limit of the theoretical understanding and the sensitivity to $\mathscr{CPT}$-relevant values of electron and proton.

- The 1S–2S frequency for hydrogen and antihydrogen (cf. Fig. 20, left).
  The 1S–2S transition energy is primarily determined by the electron or positron Rydberg constant, as this is directly proportional to the reduced electron–proton (positron–antiproton) mass. Thus (as Fig. 19 illustrates), the positron mass determines the first significant figure of ν_{1S-2S} for antihydrogen, while the antiproton mass only begins to take effect at the fourth place. The *theoretical* uncertainty for the hydrogen atom is in the eleventh place [67, 68] and is due to uncertainty of the experimental knowledge of the proton radius. Its value is determined from fits to measured elastic electron–proton scattering data, which involve extrapolations to zero momentum transfer q. For some time the most precise value of $\sqrt{<R_p^2>} = 0.862 \pm 0.012$ fm [77] was derived from data taken in Mainz at low q. A recent reanalysis of all world data, taking into account the effect of Coulomb distortion, has resulted in a significantly higher value of $\sqrt{<R_p^2>} = 0.895 \pm 0.018$ fm [78]. This value is in agreement with the value extracted from the 1S–2S laser spectroscopy under the assumption that other QED calculations are correct ($\sqrt{<R_p^2>} = 0.890 \pm 0.014$ fm [79]). In this sense, eleventh place precision in determining the hydrogen and antihydrogen 1S–2S

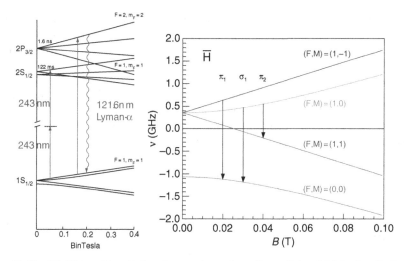

Fig. 20 The 1S–2S transition (*left*) and ground-state hyperfine splitting (*right*) of antihydrogen atoms in a magnetic field. The transitions to be measured are indicated by arrows. *Left*: the Lyman-α transition will be used for laser cooling of trapped antihydrogen

energies yields primarily information on the equality of the proton and antiproton charge distributions.

- The classical Lamb shift for hydrogen and antihydrogen.
  The frequency interval v_{2S-2P} between the $2S_{1/2}$ and $2P_{1/2}$ states of hydrogen has played an important role in the development of quantum electrodynamics since it exclusively originates from QED effects. The experimental precision for v_{2S-2P} is however rather limited due to the short lifetime of the $2P_{1/2}$ state of $\tau_{2P_{1/2}} = 1.6$ ns corresponding to a natural linewidth of 100 MHz, so that its use for high-precision $\mathscr{C}\mathscr{P}\mathscr{T}$ tests is not very promising.
- The hyperfine frequency for hydrogen and antihydrogen (cf. Fig. 20, right).
  The 1S ground state of hydrogen is split due to the interaction of electron spin S_e and proton spin S_p according to $F = S_e + S_p$ with quantum numbers $F = 0, 1$ (total spin) and $M = -1, 0, 1$ (projection of F onto the magnetic field axis). The hyperfine splitting between the $F = 0$ and $F = 1$ states of the hydrogen and antihydrogen atoms is directly proportional to both the electron(positron) and proton(antiproton) spin magnetic moments. As with the 1S–2S transition, it is extremely well known empirically for hydrogen. The impact of this on quantum physics at every stage of its development has been considerable, as Ramsey's extremely useful and informative review [80] demonstrates. These studies date back to the early 1930s, when Rabi [81, 82, 83] made a simple Stern–Gerlach beam line of inhomogeneous magnetic fields used as spin-state selectors, through which hydrogen atoms were transported. The advent of magnetic resonance methods saw determinations of the hyperfine splitting of the hydrogen ground state via microwave-induced transitions, first by Nafe and Nelson [84] and later by Prodell and Kusch [85], who reached the highest relative precision of 4×10^{-8}

of all atomic beam experiments. When it became possible to observe hydrogen atoms for times of order 10 s in a maser cavity, the precision increased accordingly [86], and it is in such experiments that the best value of ν_{HF} with relative accuracy $<10^{12}$ (cf. Table 1) was obtained.

From this it is evident that the measurement 1S–2S transitions and the ground-state hyperfine structure (GS-HFS) are complementary as they address different interactions and properties of proton/antiproton and electron/positron. Indeed, both experiments are currently being pursued at the AD of CERN.

1S–2S laser spectroscopy of antihydrogen (cf. Fig. 20, left) is the goal of the ATRAP [87] and ATHENA [88] (now transformed into ALPHA [89]) collaborations. To achieve this, they form antihydrogen from trapped antiprotons and positrons, then trap the antihydrogen atoms in a neutral-atom trap, and then perform laser spectroscopy. The trapping of neutral antihydrogen atoms requires extremely cold antihydrogen, since neutral-atoms traps have depths of $T < 1$ K ($E < 90\,\mu$eV). This has not yet been achieved (see next section).

The GS-HFS measurement is pursued by the ASACUSA collaboration [90]. Here, antihydrogen atoms escaping from a formation trap (cf. Fig. 21, left) are examined in an atomic beam line where the magnetic field gradient of sextupole magnets is used to separate the spin states of "low-field seekers" and "high-field seekers" (cf. Fig. 20, right). In this configuration, the experimental resolution is determined by the time-of-flight distribution inside the microwave cavity placed between the two sextupoles. A simulated spectrum originating from $\overline{H}$ atoms with a Maxwell–Boltzmann distribution corresponding to a temperature of 50 K is shown in Fig. 20, right; the width corresponds to a FWHM of the resonance curve of a

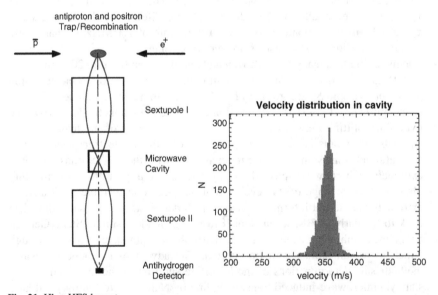

Fig. 21 Hbar HFS layout

few ppm, so that – with enough statistics – resolutions of the order of 10^{-7} can be reached. The simulations further show [91, 92] that about 100 $\overline{H}$ atoms/s in the ground state at temperatures up to 100 K produced into the full 4π solid angle are enough to perform a scan within a few days of running.

An improvement of the GS-HFS measurement will be possible once antihydrogen can be trapped and laser cooled (see Sect. 4.3.3). Microwave spectroscopy of trapped atoms has so far not reached a reasonable precision due to the thermal motion of even laser-cooled atoms in the in-homogeneous fields of the traps [93]. A possible way will be the usage of an atomic fountain setup like it was achieved for sodium atoms [94].

4.3.2 Current Challenge: Antihydrogen Production

The production of antihydrogen atoms from antiprotons and positrons in a *nested Penning trap* (see Fig. 22) has been announced by both ATHENA [95] and ATRAP [96] in 2002. The method and status have been recently reviewed in detail elsewhere [97]. While the production of several $\overline{H}$ atoms per second is reported, the measured quantum state ($n \gtrsim 50$ [98]) and temperature ($T \lesssim 2000$ K [99]) are not suitable for precision spectroscopy, although recent interpretations hint that some slower and deeper bound $\overline{H}$ might be formed [100]. The ATRAP and ALPHA collaborations are currently working on improving the formation process and building new magnets where a neutral-atom trap is superimposed onto the solenoidal field needed for their Penning traps.

ATRAP has pursued a second formation method using a *double charge exchange* method and observed a first signal [101]. It is expected that $\overline{H}$ produced by this method has lower velocity.

The ASACUSA collaboration is pursuing two other methods to produce antihydrogen, which are both targeted at producing $\overline{H}$ for the ground-state hyperfine structure measurement, where no trapping of antihydrogen is required. The method using a cusp trap, however, might be able to trap $\overline{H}$ for some time. One way is to use *Paul traps* [102], a linear one for trapping $\overline{p}$ and a novel hyperbolic two-tone trap

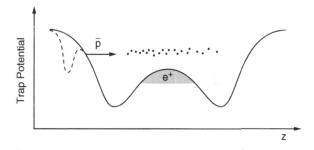

Fig. 22 Nested Penning trap. Positrons are confined in the central inverted potential, while the outer one holds antiprotons. The $\overline{p}$ traverse the electron when oscillating in the outer potential and form antihydrogen

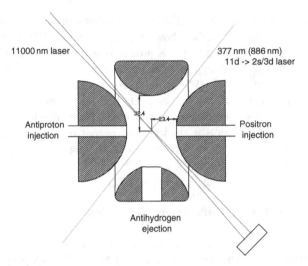

Fig. 23 Schematic view of a hyperbolic Paul trap for antihydrogen formation. Laser beams shown
are used for laser-stimulated recombination

(cf. Fig. 23) for trapping both $\bar{p}$ and e^+ for recombination. The hyperbolic trap will
use two frequencies tuned to confining both antiprotons and positrons. A critical
point is cooling of the particles, which is proposed to be done via resistive cooling
and still has to be demonstrated. A second critical procedure is the loading of the
Paul trap with $\bar{p}$ and e^+, for which the fields need to be turned on very fast. All
these techniques are now being developed. The attractive feature of a Paul trap is
that – provided the cooling works – the antihydrogen will be formed in a sphere
of less than 1 mm diameter, thus providing a point source which is important for
the following sextupole beam line to avoid aberrations that require the sextupole
dimensions to grow with the source size.

The *cusp trap* is a magnetic bottle structure known in plasma physics. It consists
of two ring currents that run in opposite directions (cf. Fig. 24 [103]). ASACUSA

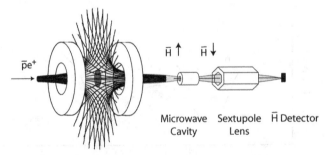

Fig. 24 Schematic view of a GS-HFS experiment using a cusp trap (*left*) which produces a polarized antihydrogen beam, a microwave cavity for spin flip, a sextupole magnet for spin analysis,
and an antihydrogen detector

has built a superconducting cusp trap and successfully confined electrons, and tests with both electrons and positrons are underway. The attractive feature of a cusp trap is that the magnetic field gradients present will separate high- and low-field seekers when they leave the trap and so the creation of a spin-polarized $\overline{H}$ beam is expected [103]. Since in cusp traps the trapping of neutral atoms has already been observed [104], the trapping of highly excited $\overline{H}$ atoms created in the trap is also expected. A recent calculation predicts that such Rydberg antihydrogen is efficiently cooled [105] leading to ground-state antihydrogen trapped in the cusp trap. This $\overline{H}$ could then be either released by lowering the magnetic field or be used for laser spectroscopy inside the trap.

4.3.3 Future Challenges: Antihydrogen Trapping and Cooling

Precision spectroscopy of the 1S–2S transition of antihydrogen as well as the above-mentioned microwave spectroscopy in an atomic fountain requires the trapping and laser cooling of neutral antihydrogen. Trapping of hydrogen and 1S–2S laser spectroscopy has been achieved first at MIT by Cesar et al. [106], but the shallow depth of Joffe-Pritchard traps of < 1 K requires the production of very cold antihydrogen. Both ALPHA [107] and ATRAP [108] at CERN-AD have now constructed new magnets where a magnetic multipole trap is superimposed onto the homogeneous magnetic field of the Penning traps. Both experiments report that charged particle clouds can be stably confined in such configurations, a fact that was heavily debated before [109, 110, 111]. So the way towards trapping antihydrogen is now open, although the temperature requirements will still make it a very stony path.

Once $\overline{H}$ is trapped, attempts will be made to laser cool it [112]. Laser cooling of hydrogen has been achieved using a pulsed Lyman-α laser [113], and recently a narrowband continuous coherent source for Lyman-α light has become available [114, 115]. Once this becomes available, the question of antimatter gravitation can be addressed [116] which has so far been impossible to investigate experimentally using charged antiparticles.

5 Summary

Low-energy antiprotons are a unique tool to study fundamental interactions and symmetries. Trapped antiprotons are very suited for precision spectroscopy of their properties as the history of achievements shows. In future, the investigation of antihydrogen offers a next step towards even higher precision for these goals.

References

1. O. Chamberlain, E. Segrè, C. Wiegand et al., Phys. Rev. **100**, 947 (1955).
2. O. Chamberlain, W.W. Chupp, G. Goldhaber et al., Phys. Rev. **101**, 909 (1956).

3. O. Chamberlain, W. Chupp, A. Ekspong et al., Phys. Rev. **102**, 921 (1956).
4. Experiment E835 at Fermilab, http://www.e835.to.infn.it/.
5. H. Koziol and S. Maury, Parameter list for the antiproton accumulator complex (aac). Technical Report, CERN, Geneva, Switzerland (1995). CERN/PS 95-15 (AR/BD).
6. G. Baur, G. Boero, S. Brauksiepe et al., Phys. Lett. B **368**, 251 (1996).
7. S. Baird et al., Design study of the antiproton decelerator. Technical Report, CERN-PS-96-043-AR, CERN, Geneva, Switzerland (1996).
8. C. Maggiore et al., Relative biological effectiveness and peripheral damage of antiproton annihilation. Proposal CERN-SPSC-2002-030; SPSC-P-324, CERN, Geneva, Switzerland (2002).
9. C. Maggiore et al., Nucl. Instrum. Meth. B **214**, 181 (2004).
10. An international accelerator facility for beams of ions and antiprotons. Conceptual Design Report, GSI (Nov. 2001).
11. PANDA Collaboration, Strong interaction studies with antiprotons. Letter of intent (2004), http://www.gsi.de/panda.
12. FLAIR – a facility for low-energy antiproton and ion research. Letter of intent (Feb. 2004), http://www.oeaw.ac.at/smi/flair/.
13. T. Baier et al., HITRAP Technical Design Report, GSI (2003), http://www.gsi.de/documents/DOC-2003-Dec-69-2.pdf.
14. G. Lüders, Ann. Phys. (NY) **2**, 1 (1957).
15. T. Lee and Y. Yang, Phys. Rev. **104**, 254 (1956).
16. C.S. Wu, E. Ambler, R.W. Hayward et al., Phys. Rev. **105**, 1413 (1957).
17. J.H. Christenson, J.W. Cronin, V.L. Fitch et al., Phys. Rev. Lett. **13**, 138 (1964).
18. B. Aubert et al., Phys. Rev. Lett. **87**, 091801 (2001).
19. K. Abe et al., Phys. Rev. Lett. **87**, 091802 (2001).
20. L. Iconomidou-Fayard and J.T. Thanh Van, eds., XIV Rencontres de Blois: Matter–Antimatter asymmetry, The Gioi Publishers, Vietnam (2002).
21. A.D. Dolgov, In L. Iconomidou-Fayard and J.T. Thanh Van, eds., XIV Rencontres de Blois: Matter–Antimatter asymmetry, The Gioi Publishers, Vietnam (2002), pp. 15–25.
22. A.D. Sakharov, JETP Lett. **5**, 24 (1967).
23. F.W. Stecker, In L. Iconomidou-Fayard and J.T. Thanh Van, eds., XIV Rencontres de Blois: Matter–Antimatter Asymmetry, 5–14, The Gioi Publishers, Vietnam (2002).
24. O. Bertolami, D. Colladay, V.A. Kostelecky et al., Phys. Lett. B **395**, 178 (1997).
25. D. Colladay and V.A. Kostelecký, Phys. Rev. D **55**, 6760 (1997).
26. R. Bluhm, V.A. Kostelecký, and N. Russell, Phys. Rev. Lett. **79**, 1432 (1997).
27. R. Bluhm, V.A. Kostelecký, and N. Russell, Phys. Rev. D **57**, 3932 (1998).
28. V.A. Kostelecký, Phys. Rev. Lett. **80**, 1818 (1998).
29. R. Bluhm, V.A. Kostelecký, and N. Russell, Phys. Rev. Lett. **82**, 2254 (1999).
30. G. Gabrielse, A. Khabbaz, D.S. Hall et al., Phys. Rev. Lett. **82**, 3198 (1999).
31. H. Dehmelt, R. Mittleman, R.S. Van Dyck, Jr. et al., Phys. Rev. Lett. **83**, 4694 (1999).
32. R.K. Mittleman, I.I. Ioannou, H.G. Dehmelt et al., Phys. Rev. Lett. **83**, 2116 (1999).
33. D. Bear, R.E. Stoner, R.L. Walsworth et al., Phys. Rev. Lett. **85**, 5038 (2000).
34. D.F. Phillips, M.A. Humphrey, E.M. Mattison et al., Phys. Rev. D **63**, 111101 (2001).
35. V.W. Hughes, M.G. Perdekamp, D. Kawall et al., Phys. Rev. Lett. **87**, 111804 (2001).
36. J.M. Link, et al., Phys. Lett. B **556**, 7 (2003).
37. W.-M. Yao, C. Amsler, D. Asner et al., J. Phys. G **33**, 1+ (2006).
38. G. Gabrielse, X. Fei, K. Helmerson et al., Phys. Rev. Lett. **57**, 2504 (1986).
39. M. Iwasaki, S.N. Nakamura, K. Shigaki et al., Phys. Rev. Lett. **67**, 1246 (1991).
40. G. Blanford, D. Christian, K. Gollwitzer et al., Phys. Rev. Lett. **80**, 3037 (1998).
41. J.K. Thompson, S. Rainvilleand, and D.E. Pritchard, Nature **430**, 58 (2004).
42. T. Yamazaki, N. Morita, R.S. Hayano et al., Phys. Rep. **366**, 183 (2002).
43. T. Yamazaki, E. Widmann, R.S. Hayano et al., Nature **361**, 238 (1993).
44. N. Morita, M. Kumakura, T. Yamazaki et al., Phys. Rev. Lett. **72**, 1180 (1994).
45. H.A. Torii, R.S. Hayano, M. Hori et al., Phys. Rev. A **59**, 223 (1999).
46. V.I. Korobov and D.D. Bakalov, Phys. Rev. Lett. **79**, 3379 (1997).

47. V.I. Korobov, In E. Zavattini, D. Bakalov, and C. Rizzo, eds., Frontier Tests of Quantum Electrodynamics and Physics of the Vacuum, Heron Press, Sofia (1998), pp. 215–221.
48. Y. Kino, M. Kamimura, and H. Kudo, Nucl. Phys. A **631**, 649c (1998).
49. Y. Kino, M. Kamimura, and H. Kudo, Innovative Computational Methods in Nuclear Many-Body Problems, Towards a New Generation of Physics in Finite Quantum Systems (1998).
50. N. Elander and E. Yarevsky, Phys. Rev. A **56**, 1855 (1997). Errata **57**, 2256 (1998).
51. M. Hori, J. Eades, R.S. Hayano et al., Phys. Rev. Lett. **91**, 123401 (2003).
52. V.I. Korobov, Phys. Rev. Lett. **67**, 62501 (2003). Erratum Phys. Rev. A **68**, 019902.
53. Y. Kino, M. Kamimura, and H. Kudo, Nucl. Instrum. Meth. Phys. Res. B **412**, 84 (2004).
54. A.M. Lombardi, W.Pirkl, and Y. Bylinsky, In P. Lucasa and S. Webber, eds., Proceedings of the 2001 Particle Physics Accelerator Conference, IEEE, Piscataway, NJ (2001), pp. 585–587.
55. M. Hori, A. Dax, J. Eades et al., Phys. Rev. Lett. **96**, 243401 (2006).
56. R. Hughes and B.I. Deutch, Phys. Rev. Lett. **69**, 578 (1992).
57. E. Widmann, J. Eades, T. Yamazaki et al., Phys. Lett. B **404**, 15 (1997).
58. D. Bakalov and V.I. Korobov, Phys. Rev. A **57**, 1662 (1998).
59. E. Widmann, J. Eades, T. Yamazaki et al., Phys. Rev. Lett. **89**, 243402 (2002).
60. V.I. Korobov and D. Bakalov, J. Phys. B **34**, L519 (2001).
61. N. Yamanaka, Y. Kino, H. Kudo et al., Phys. Rev. A **63**, 012518 (2001).
62. Y. Kino, N. Yamanaka, M. Kamimura et al., Hyper. Interact. **146–147**, 331 (2003).
63. A. Kreissl, A.D. Hancock, H. Koch et al., Z. Phys. C **37**, 557 (1988).
64. D. Bakalov and E. Widmann, Determining the antiproton magnetic moment from measurements of the hyperfine structure of antiprotonic helium (2007), http://arxiv.org/abs/physics/0612021.
65. M. Charlton, J. Eades, D. Horváth et al., Phys. Rep. **241**, 65 (2004).
66. M. Fischer, N. Kolachevsky, M. Zimmermann et al., Phys. Rev. Lett. **92**, 230802 (2004).
67. J.R. Sapirstein and D.R. Yennie, In T. Kinoshita, ed., Quantum Electrodynamics, World Scientific, Singapore (1990), pp. 560–672.
68. K. Pachucki, Private communication (2003)..
69. S.R. Lundeen and F.M. Pipkin, Phys. Rev. Lett. **46**, 232 (1981).
70. K. Pachucki, D. Leibfried, M. Weitz et al., J. Phys. B: At. Mol. Opt. Phys. **29**, 177 (1997).
71. S.G. Karshenboim, In S.G. Karshenboim and V.B. Smirnov, eds., Precision Physics of Simple Atomic Systems, Springer, Berlin, Heidelberg (2003), pp. 142–162. Hep-ph/0305205.
72. S.G. Karshenboim, Phys. Lett A **225**, 97 (1997).
73. M.S. Fee et al., Phys. Rev. A **48**, 192 (1993).
74. K. Hagiwara et al., Phys. Rev. D **66**, 010001 (2002).
75. R.S. Van Dyck, P.B. Swinberg, and G. Dehmelt, Phys. Rev. Lett. **59**, 26 (1987).
76. M. Niering, R. Holzwarth, J. Reichert et al., Phys. Rev. Lett. **84**, 5496 (2000).
77. G.G. Simon, C. Schmitt, F. Borokowski et al., Nucl. Phys. A **333**, 381 (1980).
78. I. Sick, Phys. Lett. B **476**, 62 (2003).
79. T. Udem, A. Huber, B. Gross et al., Phys. Rev. Lett. **79**, 2646 (1997).
80. N. Ramsey, In T. Kinoshita, ed., Quantum Electrodynamics, World Scientific, Singapore (1990), pp. 673–695.
81. I.I. Rabi, J.M.B. Kellogg, and J.R. Zacharias, Phys. Rev. **46**, 157 (1934).
82. I.I. Rabi, J.M.B. Kellogg, and J.R. Zacharias, Phys. Rev. **46**, 163 (1934).
83. J.M.B. Kellogg, I.I. Rabi, and J.R. Zacharias, Phys. Rev. **50**, 472 (1936).
84. J.E. Nafe and E.B. Nelson, Phys. Rev. **73**, 718 (1948).
85. A.G. Prodell and P. Kusch, Phys. Rev. **88**, 184 (1952).
86. H.M. Goldenberg, D. Kleppner, and N.F. Ramsey, Phys. Rev. Lett. **8**, 361 (1960).
87. ATRAP collaboration, http://hussle.harvard.edu/~atrap.
88. ATHENA collaboration, http://cern.ch/athena.
89. ALPHA collaboration, http://alpha.web.cern.ch/alpha.
90. ASACUSA collaboration, http://cern.ch/asacusa.
91. E. Widmann, J. Eades, R. Hayano et al., In S.G. Karshenboim, F.S. Pavone, F. Bassani, et al., eds., The Hydrogen Atom: Precision Physics of Simple Atomic Systems, Springer, Berlin, Heidelberg (2001), pp. 528–542, arXiv:nucl-ex/0102002..
92. B. Juhasz, D. Barna, J. Eades et al., In D. Grzonka, R. Czyzykiewicz, W. Oelert, et al., eds., Proceedings of LEAP03, vol. 796, AIP (2005), pp. 243–246.

93. A.G. Martin, K. Helmerson, V.S. Bagnato et al., Phys. Rev. Lett. **61**, 2431 (1988).
94. M.A. Kasevich, E. Riis, S. Chu et al., Phys. Rev. Lett. **63**, 612 (1989).
95. M. Amoretti, C. Amsler, G. Bonomi et al., Nature **419**, 456 (2002).
96. G. Gabrielse, N.S. Bowden, P. Oxley et al., Phys. Rev. Lett. **89**, 213401 (2002).
97. G. Gabrielse, Adv. At. Mol. Opt. Phys. **50** (2005).
98. G. Gabrielse, N.S. Bowden, P. Oxley et al., Phys. Rev. Lett. **89**, 233401 (2002).
99. G. Gabrielse, et al., Phys. Rev. Lett. **93**, 073401 (2004).
100. T. Pohl, H.R. Sadeghpour, and G. Gabrielse, Phys. Rev. Lett. **97**, 143401 (2006).
101. C.H. Storry, A. Speck1, D.L. Sage et al., Phys. Rev. Lett. **93**, 263401 (2004).
102. ASACUSA Collaboration, Atomic spectroscopy and collisions using slow antiprotons. Proposal CERN-SPSC-2005-002, SPSCP-307 Add.1, CERN, Geneva, Switzerland (2005).
103. A. Mohri and Y. Yamazaki, Europhys. Lett. **63**, 207 (2003).
104. A.L. Migdall et al., Phys. Rev. Lett **54**, 2596 (1985).
105. T. Pohl, H.R. Sadeghpour, Y. Nagata et al., Phys. Rev. Lett. **97**, 213001 (2006).
106. C.L. Cesar, D.G. Fried, T.C. Killian et al., Phys. Rev. Lett. **77**, 255 (1996).
107. G. Andresen, W. Bertsche, A. Boston et al., Phys. Rev. Lett. **98**, 023402 (2007).
108. G. Gabrielse, P. Larochelle, D.L. Sage et al., Phys. Rev. Lett. **98**, 113002 (2007).
109. T.M. Squires, P. Yesley, and G. Gabrielse, Phys. Rev. Lett. **86**, 5266 (2001).
110. E.P. Gilson and J. Fajans, Phys. Rev. Lett. **90**, 015001 (2003).
111. J. Fajans, W. Bertsche, K. Burke et al., Phys. Rev. Lett. **95**, 155001 (2005).
112. J. Walz, Phys. Scripta **70**, C30 (2004).
113. I.D. Setija, H.G.C. Werij, O.J. Luiten et al., Phys. Rev. Lett. **70**, 2257 (1993).
114. K. Eikema, J. Walz, and T. Hänsch, Phys. Rev. Lett. **83**, 3828 (1999).
115. K. Eikema, J. Walz, and T. Hänsch, Phys. Rev. Lett. **86**, 5679 (2001).
116. J. Walz, and T.W. Hänsch, General Relativity Gravitation **36**, 561 (2004).

Index